Other Books by Tim Coonan:

Decline and Recovery of the Island Fox: A Case Study for Population Recovery (Cambridge University Press, 2010).

Our Lady, Queen of the Highways (Adelaide Books, 2022)

Lake Effect

Tim Coonan

Deer Run Press
Waldoboro, Maine

Copyright © 2023 Tim Coonan.

All rights reserved. No part this book may be reproduced or transmitted in any form or by any means, electronic or mechanical, including photo-copying, recording, or by any information storage and retrieval system, without written permission from the copyright owner.

Library of Congress Card Number: 2023946021

ISBN: 978-1-937869-22-9

First Printing, 2023

Published by
Deer Run Press
8 Cushing Road
Cushing, ME 04563

Contents

1 The Smell of Summer 1
2 A Fur Piece 4
3 Joe Hefty 8
4 Coonan Cove 11
5 A Trusty Steed, and a Record-breaking Catch. And Tiny Deer 14
6 Holy Communion and Hot Fudge 17
7 The Volcano on the Horizon 19
8 Rotten Eggs 23
9 Whiskey's for Drinking 29
10 The Gold Standard. Or Shangri-la 40
11 The Return of the King 49
12 The Russians Are Coming 55
13 Morning on the Deck 63
14 Biological Diversity 74
15 Big, Fierce Things 81
16 Dock Talk 88
17 Dinner Will Be Late 96
18 The Smoke from a Distant Fire 103
19 Out of the Frying Pan 111
20 Is it Hot in Here, or Is it Just Me? 117
21 The Otter, the Nutcracker and the Wolf 126

Dedication

*For my dad, Tom Coonan,
whose attention was caught
by a Lake Almanor brochure in 1971.*

*And for Terry, Dan, Katie, Nora and Dennis,
who, because of that, discovered the pleasures of a lake
vacation among the tall pines on Lake Almanor's shores.*

Acknowledgements

It's all a journey, isn't it? We get to where we were going, or somewhere close to that, or even someplace completely different from where we thought we were headed. But we have companions along the way. My companion in this writing journey of mine, and in life itself, is Nell, my partner, and I thank her for her support, her patience and her critical eye. Thank goodness she likes Lake Almanor!

My brother Dan has encouraged me on this journey (indeed, he is on his own) and he figures largely in this telling. His affinity for (shall we say obsession with?) Lake Almanor is really why we all have met at the lake in recent years. He is the straw that stirs the family drink, as I like to say. I thank Dan for reviewing this, and for adding his own memories to mine (which are perhaps a little murky, like lake water). I also thank my friend Dan Dutcher for reviewing the manuscript, and Nancy Parsons did a much-needed job of editing it; she had to deal with, among other things, my penchant for using semicolons. Carrie Coonan, wildlife photographer and budding biologist, provided some great photos of Almanor and its denizens. And my dad—well, Dad started this whole darned Almanor thing. Thank you!

Foreword

Parents are required to take their kids on family vacations. It's pretty much the law, in every state, no getting around it. These family vacations can take a multitude of forms, of course, and they are all acceptable, under the law. You could go camping (we didn't). You could go hiking (we didn't, except for once, as you'll see). Fishing? Nope. Snow skiing, water skiing: again, no. Boat cruises? Not on your life, and I still can't imagine doing that. Brings up images of the Irish immigrants on the *Titanic*, stuck down in steerage. Sure, the parties were fun, until the water started to rise. But road trips? You bet. With a purpose, of course, to visit relatives in the Midwest and East, where our family roots were; we were the family members who had moved out west, away from the family clans. I think everyone thought my parents a bit crazy, somewhat weird for doing that. And I don't think Mom EVER got over the fact that Dad had taken her away from her beloved Illinois.

In addition to family station wagon road trips, we did lake cabin vacations. Classic, aren't they? A rustic wood cabin, on the shore of a lake lined with pine trees. Clear, quiet nights. No city lights to dim the majesty of the night sky, more stars than you've ever seen, even the magnificent Milky Way, our home galaxy, appearing overhead. Cold, or at least cool water to swim in, maybe off a wooden dock. Fishing from the same dock. Maybe a loon or two out on the water. What's not to like? The lake cabin has certainly been a tried-and-true vacation genre in North America, the setting for numerous books and movies (along with its offshoot, the lake summer camp), and even the occasional horror flick. And it's so much easier, and neater, than camping! Sleeping in a bed under a roof, safe from mosquitoes and gnats and ticks and wood smoke and whatever wild creature just rubbed up against the side of the tent. Not tracking dirt into the tent, and being able to wash that same dirt off your feet in an indoor shower. Not having to struggle out of your sleeping bag and tent, grab a flashlight, head to the campground restroom, just to pee in the middle of the night. Yes, cabins have all the creature comforts we want, if maybe to a lesser degree than our suburban homes. But with a far better view.

So we took lake cabin vacations. Now, I don't believe my dad's family had ever done this type of thing; he grew up in Maryland, and I've heard no stories of his professor father taking the family on any lake vacations. But my mom's family did, in fact, do the lake cabin family vacation thing. She grew up in Springfield, Illinois, and her dad, an avid outdoorsman, would drive them north, deep into upper Wisconsin, to a lake with decidedly cold water. Being an experienced angler, I'm sure my grandfather loaded the car with their long bamboo fishing poles (with which we'd fish on Lake Springfield, decades later) and my grandmother likely cleaned lots of fish for their cabin dinners. At that lake, Devil's Lake, they had what I consider real fish: walleye and northern pike. On our own lake vacations, my mom didn't have to clean any fish for dinner, because we didn't catch any (with one exception, as, again, you'll see).

It occurs to me that much of our lives, if not our collective history and even evolution itself, hinges on somewhat random events, takes a turn, veers one way rather than another, and then stays that course. Proto-humans stand upright in the African savannah, and millennia later a human runs a sub-four-minute mile. Reptilian *Archaeopteryx* glides onto a tree branch with these newfangled feathers, and eventually an osprey deftly plucks a trout from lake waters. So it was with these lake vacations. We ended up at Lake Almanor, rather than Lake Havasu or Bass Lake or Lake Tahoe or Shasta Lake, and we stayed the course.

This, itself, is not remarkable in any historic or significant sense. It's just what we did. And still do, after a decades-long hiatus. Is that a homecoming of sorts? I suppose so. But you can't read too much into it, really. Many families do this, return to favorite haunts. Sure, it's a refuge, in a way, for me, and certainly for my brother Dan, and for my dad, who found the lake comforting to return to after my mom's passing; it was, after all, the scene of many great family experiences. But I've also realized that, because our time at the lake spans, or at least brackets, a 50-year time period, the lake is a frame of reference or a lens on the environmental changes that have occurred during that time. You can't help but notice them. The lake and our experience of it are not immutable. We've had to adjust our expectations of our lake experience, for one, but it also causes me to reflect on the changes, already irreversible and regrettable, caused by climate change, and evident in our lake visits. I suppose this is, then, an account of what was.

Aw, what the hell? It's still pretty up there. And I like pine trees.

Lake Effect

Chapter 1

The Smell of Summer

"The trees are so tall and straight," Dad observed. "How do they grow so straight?" We were driving away from Lake Almanor, among the white firs and ponderosa pines, the occasional massive sugar pine, with its ridiculously oversized cones (how could any squirrel resist that super-stimulus?). Dad was 86, and had grown forgetful—but only of recent events and circumstances; his long-term memory was in full gear and active. And he retained a childlike sense of wonder, of keen interest, about the natural world. Some of this stemmed from his background as a scientist, his long career (38 years) as a chemical engineer for Standard Oil/Chevron. A lifelong interest in science, fed further by his "second career": fifteen years as a middle school science teacher. He had never taken biology courses when in school himself, and he had thoroughly enjoyed the challenge of understanding life science enough to teach it to seventh graders. But he still relied on me, a wildlife biologist, to answer biological questions. Such as why conifer trees grew so straight and tall.

Now, I'm no botanist (which is why, as a wildlife biologist, I hired botanists to do botanist stuff), but I told him evolution was certainly at play here: it was a race to the top for these trees, a race to the light that was at the top of the crowded conifer forest. Light was everything. If you want to photosynthesize, you need that light. As I tell MY seventh graders (yes, I, too, have a second career as a middle school science teacher), the sun's light is the ultimate source of energy here

Tim Coonan

on earth. So evolution placed a premium on getting to that light, and conifers grow tall and straight, at least in these crowded mixed conifer forests, full of competitors, in order to get up there. No doubt aided by certain plant hormones, but I didn't get into that. We'd stick with ultimate causes for now.

Dad thoroughly enjoyed the drive through the tall trees, both to and from Lake Almanor, and so did I. It had been fifty years since Dad first took our family to the lake, in 1971. What began as a typical family lake vacation turned into something else entirely. Our return trips (not annual by any means, with gaps even of decades) turned Almanor into a sacred family place. A happy place. We all have those. And for me, especially—as well as for my brother Dan—Almanor is one of those happy places. Almanor, for me, taps into my love of earth's natural beauty, and its biological richness and diversity. And trees.

I love conifers, in particular, the ponderosa pine. Now, I don't really know why this is so. Why does ANYONE take a shine to one thing or another, hold it in high regard, accord it status? Can't say. But ponderosas do it for me. Flagstaff, where I did my master's work, was surrounded by ponderosas, and that's where I attached nestboxes to ponderosas in order to attract and study American kestrels (North America's smallest falcon). And I HAVE to see big trees at least once a year. Lately this has been the annual trip to Lake Almanor.

One thing I love is that a ponderosa pine forest is the smell of summer, for me. Now, there are many things that smell of summer. Salt air. Coconut-scented sunscreen. Campfires. Even insect repellant—you know you're somewhere slightly wild when you smell DEET. But there's nothing like the smell of a pine forest in the summer. The sun's rays, filtered through the pine needle canopy, warm the boles of the trees enough for them to release terpenes. Being a chemical engineer, and cracking hydrocarbons all his career, my dad knows what terpenes are: they're hydrocarbons that, when volatilized (gassed up) smell pretty good. Like fruity, heady aromas. Stick your nose in a furrow between bark pieces of a pine tree in the summer, and breathe in. You'll smell butterscotch or vanilla, maybe even something more exotic. And on a really warm summer day in a pine forest, you don't even need to stick your nose in the bark; the faint smell wafts through the forest itself. It's the smell of summer in the high country, and I'm somewhat addicted to it. It's one thing I look forward to on our now-pretty-much-annual Almanor trip.

However, things change, as much as I would like them to stay the same, and the change is not always for the better. These days the smell of the high country is something other than pine trees releasing butterscotch and vanilla terpenes. It's now the smell of pine trees burning. It's the smell of wildfire.

Pines and firs tower above a meadow's edge at Lake Almanor.

Chapter 2

A Fur Piece

Dad doesn't remember why exactly or how he chose Lake Almanor as a lake vacation spot for us. His memory, even his long term one, is pretty selective; he thinks he may have seen an ad for Lake Almanor in a brochure or magazine. The thing is, it was a long way, over five hundred miles, from our southern California home. There are closer lakes, such as Bass Lake, which we actually went to in 1968. There we fished and swam; Bass Lake has a big rock which swimmers slide down, and we did that, too. You can see us doing that on the 8mm home movies my dad took. Bass Lake is in the southern Sierra Nevada, and pretty close to Yosemite, which we visited on that trip; there are pictures of us at Half Dome, and Inspiration Point, and among the big trees at Wawona. We boys are in Chuck Taylors or Keds, jeans and long-sleeved plaid shirts (this was before I discovered the joy of shorts and t-shirts).

I distinctly remember a wildlife encounter I had there. Walking near the developed area at Wawona, I saw a spotted fawn, a very young one, lying in the grass near the trail. It didn't move at all, and I was amazed at its presence: how close we were to it, the fact that it didn't move. It was beautiful.

Bass Lake was smaller, hotter, and more crowded, but also closer to our LA home. Why didn't we go back there? My brother Terry suggests, perhaps jokingly, it was because of the Hell's Angels presence. For many years, Bass Lake was the site of an annual Hell's Angels rally, and if that wouldn't

Lake Effect

deter a mild-mannered Catholic family from visiting, I don't know what would.

In 1969, my dad planned a family vacation to June Lake, in the Mammoth area of California, on the east side of the Sierra Nevada. He even booked a week at a cabin up there. But Dad's sister, Pat, decided to marry John that summer, before they were deployed to Germany with the army, and so the June Lake plans were scrapped. Instead, we embarked on our second family cross-country trip, from southern California to Massachusetts, to see Pat marry John. We never made it back to June Lake (I did make it there as an adult. Cathy and I took the kids there for a lake/fishing vacation—during which we caught NO FISH whatsoever. Powerbait, my ass). One thing that occurs to me is that if Pat hadn't married John—which is something else to consider entirely—we might be returning to June Lake every year, instead of Lake Almanor.

The trip to Lake Almanor was so long that we took two days to do it; we stayed overnight in Sacramento, so we could arrive at Almanor mid-day. Beyond Sacramento, we exited the main artery, I-5, onto the more secondary roads leading to Almanor. The lake is a bit off the beaten path, which is part of its charm, its allure. Lake Shasta, to the west, is pretty much adjacent to I-5 and chock full of houseboats and ski boats. Tahoe, to the east, is a summer and winter mecca, with ski resorts and even lakeside casinos. It's a bit much. Plus, it's deep and it's cold—you can't swim in Lake Tahoe! Almanor is between the two, and along no major highway. You pretty much have to want to go to Lake Almanor. To get there, you take highway 99 or 70 toward Oroville, and Chico, through towns like Marysville, which are pretty cute. I still like to drive through Marysville—who wouldn't want to live in a town named that? Old, cute main street, stately old brick high school set off in a field. I could see myself teaching there. From Marysville the road runs north through farmlands and orchards, stonefruits in particular. In fact, on the drive back from Almanor, my parents would stop at a roadside stand and buy a flat of plums. My dad says the car smelled like plums all the way home. In the back of the station wagon, we'd help ourselves to the ripe fruit.

The Feather River drains a big swath of the Sierra/Cascade juncture (and has great swimming holes, as Carrie can attest).

Lake Effect

There are two routes into Almanor, and often my dad would choose the lower one, which arrives at the bottom of the heart-shaped lake by way of the Feather River Canyon, a beautiful, precipitous gorge studded with tall timber. The road follows the bottom of the canyon, with ever-changing views of the boulder-lined river, occupied alternately by narrow rapids and by wider, quiet pools. Freight trains of the Western Pacific Railroad chug deliberately and regularly through the canyon, increasing its allure (at least to me, a railroad buff).

The upper route climbs out of Chico, a hot, tree-lined college town, up a dry mesa. That gradually gives way to sparse conifers set among volcanic-rock strewn fields, before diving down lush Deer Creek to Lake Almanor. Here's where the tall trees, which my dad so enjoyed driving through, become thick, and the riparian willows and alders overhang the small winding creek, frequented by anglers. As you approach the bottom of Deer Creek the forest occasionally opens up into lush meadows bordered by old split-rail fences, hinting of former homesteads and high-country ranches. You're almost there. Just a few more miles on Highway 36, through big trees and past weathered cabins and a few old motor lodges —but why stay there when it was a just a few more miles to the lake? The lake was everything. Otherwise, it would just be a cabin in the woods.

Chapter 3

Joe Hefty

Water was a required element for a family vacation. I'm sure as soon as we checked into the cabins—the Knotty Pine Lodge—we grabbed our towels and headed down the hill to the swimming area, near the docks in Big Cove. The lodge was standard fare, and classic. Individual or duplex wood cabins, made completely of pine, a wood that absorbs sunlight and adopts a rich hue of yellow as it ages. Terry was likely the first of us in the water. My twin has always been bigger than me, bolder, more confident, brazen, even. Dan was three years younger and not far behind Terry in those qualities. We had a continual competition: the first into the water, be it lake or pool, was "Joe Hefty," a testosterone-laden daily title.

The rustic classic Knotty Pine Lodge just screamed "lake vacation."

Lake Effect

The lake was the main event for us kids. Everybody swam, except for Dennis, who was barely over a year old at that point. My mom, who was of the bathing cap era, made the girls wear bathing caps, even at the lake. Now, I can see that bathing caps make Olympic swimmers sleeker and faster, but I fail to see their utility, otherwise. I mean, they certainly don't keep your hair dry if you're dunking your head underwater. No matter. The pictures don't lie. In 1971, the girls in our family wore bathing caps. My dad was a good, strong swimmer himself. When he took us to free swim at our hometown pool (the "Plunge"), he'd swim laps while we played Marco Polo and corner tag and dove off the low and high dives (well, we JUMPED off the high dive, at least). When we returned home from free swim Mom would fix popcorn, served in individual colored metal bowls, at the bottom of which were salt and unpopped kernels, when the popcorn was gone.

If all we had done at the lake was swim, that would have been just fine with us.

Tim Coonan

We swam every day, sometimes twice or more. The towels which hung on the railing and the bathing suits in the bathroom never really dried before they were pressed into service again. We swam from the docks or the shore of the cove, which also had a wooden diving or swimming platform anchored not far offshore, on which we spent considerable time, racing out there and hanging out. Wooden dive platforms are long gone, the wooden planks alone posing a liability not tolerated these days. There were also small kayaks at the docks for our use, and we fooled around in those, in the back end of the cove. That felt pretty independent, as I recall.

At the beginning of the week, your lake vacation stretches out before you. It might even seem a bit long. But this wasn't a month or a summer (summer at the shore, or the Cape, maybe?), it was just a week. So by Wednesday, you realized the week was half over, and the potential of the days ahead started to diminish; by Friday you just wanted to wring everything you possibly could out of what was left of your week at the lake, and that meant swimming as much as possible. Even though your parents had planned other activities.

Chapter 4

Coonan Cove

We were not boat people, in any sense of the term. No ski boat on a trailer in the driveway. No sailboat at the marina. Now, my dad did learn to sail as a kid; his dad was a professor at the Naval Academy and my dad took lessons there. But our only boating experience was on the patio boats at Lake Almanor. They could be rented right from Big Cove, from the Knotty Pine, even. They were, and still are, flat-bottomed floaty boats with patio covers, perfect for a family outing on a flat, windless lake. And unsuitable for pretty much anything else. The basic design has changed little in decades, except for the addition of cupholders and a speaker system. And more comfortable seats.

It was one activity we didn't mind, in fact we even looked forward to it, because it involved swimming, and, moreover, we got to drive the boat. Mom packed the Styrofoam cooler with lunch, and with baseball hats, towels and sunscreen, we'd motor slowly (the only speed possible) out of Big Cove and down the peninsula. I don't recall that many ski boats on Lake Almanor back in the day, though these days there are more, along with jet skis. It's not that we had the lake to ourselves, but it sure felt like that. We'd pass a few fishermen trolling the shores for bass, as Dad motored down pass the tip of the peninsula to the far side of the lake, where we'd find a quiet cove and anchor up. Then it was swimming time, in what seemed to us to be our own private cove. In fact, we ended up calling these places "Coonan Cove," as if we had discovered it, as if no one, native or colonist, had ever

Tim Coonan

anchored up there or otherwise laid claim to it. The illusion, or at least tongue-in-cheek conceit, of discovery, or ownership. No one else would ever have this same sense of discovery and land (water?) claim as we did, that day on that spot.

Lunch was the typical family picnic-vacation lunch we were used to: sandwiches, chips, fruit, likely cookies, and soda (odds are Orange Crush on the latter). Then, even though our time on the boat, for which we had paid dearly, was ticking away, we HAD TO WAIT AN HOUR UNTIL SWIMMING AGAIN. Fear of the dreaded cramps drove this 1960s practice, pretty much written into law and followed faithfully by families of all creed and color, it being one of the most adhered-to health assumptions ever conceived by humans. It's funny to look back and realize that THAT was what people were worried about – not smoking, not fat, sugar or carbohydrate intake, not red meat, not lack of exercise or too much sitting. But God help you if you went in swimming right after eating.

Eventually, we did go back in swimming, and I don't recall any of us getting cramps. And then we'd haul in the anchor—a plastic bottle filled with sand—and turn back toward Big Cove, which is when the real fun began: we kids got to drive the boat, under Dad's watchful eye, of course. Even the little ones got in on the action, sitting on Dad's lap to do this. But boy were we big shots, driving that boat. What a sign of maturity. It would be years before we could get our licenses, drive a car, but by golly we had what it took to drive a boat (which tells you how poorly regulated boat driving is). Photos of this were required, mandatory. And so there are photos of every one of us, some serious, some casually playing it off ("Yeah, I do this all the time") in our baseball caps and Little League shirts, bent over the wheel of the boat.

Boats are relaxing, the drone of the engine and the gentle surge and slip-slap of the boat lulling some of us to sleep on the way back. Dennis asleep on Mom's lap, in his puffy orange life preserver (which would have made him bob like a cork, like a red-and-white fishing bob, had he fallen in). Pulling back into the dock, through crowded Big Cove, was perhaps the diciest part of the trip, and Dad took over, guided in by the tanned, surly teens working the boat dock. We disembarked, trudged back up to the cabin. And then we went swimming again.

One of several "Coonan Coves" discovered over the years, each never visited by humans until we arrived.

We showed prowess and promise as high-speed boat drivers, thanks to our turns at the wheel of the sleek patio boat.

Chapter 5

A Trusty Steed, and a Record-breaking Catch. And Tiny Deer

Now, for some reason, horseback riding figures strongly in summer lake vacations. Kids love horses, or the idea of them, the idea of riding horses. Sure, my sisters collected those plastic horses, of different colors and breeds. Horses ARE beautiful animals, I'll admit. The attraction to girls is understandable. My daughter Carrie just wanted to be around horses, work in their stalls, etc., but was told by stable owners she'd have to take lessons first. So she did, or more accurately, we did, over several years, at several different stables in the Ventura area. At one, she had to go catch the horse in a field first, and that was harder that it might seem. I don't really think horses want to be ridden, all that much. Would you enjoy hiking with a bridle in your mouth, and a heavy saddle cinched tight around your middle? Carrie enjoyed riding for several years, and there were lots of choices to be made (Western versus English, etc.). Until she was thrown, as her horse balked when Carrie attempted a small jump. I was there, and I hated seeing that. They're huge animals, and you're up pretty high....

In any case, we went horseback riding several times at Almanor, on the west side of the lake. Admittedly, it was pretty cool to be on top of a horse on a trail above the lake. And of course, it was all about which horse you got, and its name (none of which I now remember). One time Nora was bucked off her horse, even hit a tree root on her way down. She says it was traumatic, so much so that she had repressed the memory until I brought it up. I had to lead her horse back to the stable. On that same ride, the stable owner asked Mom which of us had the most experience on a horse. Katie, she said. Turned out they were a little short on horses that day. Katie ended up riding a burro.

Horseback riding was hot, dusty, and mandatory. Fishing was not—but I was rewarded with a record-braking catch.

Tim Coonan

Horseback riding, like swimming and the patio boat, was a required element for us; we went riding each of the three trips we made to Almanor. And it's funny how that holds no allure for me now, or to almost any adult (who isn't already a horse person). Who would want their arthritic back and thin butt tossed around by the piston-like movement of a horse's back?

Fishing appealed to me, perhaps more than anyone else, and so I did a little fishing off the dock at Knotty Pine. Nothing serious. Likely with minnows and a red-and-white bobber. But with record-setting results: the tiniest fish ever caught at Lake Almanor. Even more notably, it was one of perhaps only three fish I have ever caught in my life. Making it all the more significant.

Being suburban kids from LA, we were also fascinated with the fairly tame mule deer that frequented the peninsula, helping themselves to the well-watered lawns and ornamentals of the peninsula houses. The deer were commonly seen, and unafraid; there were no predators, or hunting, on the peninsula. It was mainly does with young of the year, slightly older fawns, with the occasional spike buck or older buck, antlers still in velvet during the summer season. This was our closest brush with "wildlife", and it might as well have been on Mutual of Omaha's Wild Kingdom, as far as we were concerned. This was the edge of the wild for us, and we wanted to come back with award-winning wildlife photos. Thus, we took pictures of fairly far-off deer with our Kodak Instamatics and fully expected that they could be blown up into dramatic 8x10 enlargements at our corner drugstore back in El Segundo. Sadly, that was not the case, and we ended up learning our first lessons about the limits of photographic resolution. But even to this day, I still take pictures of the deer up there.

Chapter 6

Holy Communion and Hot Fudge

National parks (with which I am thoroughly familiar, having been a biologist for the National Park Service for 30 years) all have gateway communities, small cities which serve as portals to the magnificent national parks outside their town limits. These vary from gaudy and over-commercialized spots like Gatlinburg, Tennessee, where wax museums and Dollywood usher you into the Great Smoky Mountains, to charming, well-planned communities like Springdale, Utah, which transition one somewhat more seamlessly into majestic Zion National Park. If Lake Almanor were a national park —which it decidedly is not—its gateway community would be tiny Chester, located at the top of the lake (atop the left atrium, if you're following the heart analogy).

Chester began life in the late 19th century as merely a stop on the stage road from Red Bluff to Susanville, gradually acquiring first a post office and then a rural school. The city is spread along Highway 36, and parts of it are noticeably old. One section, at the eastern end of town, before you cross the marsh and head up the hill toward the head of the peninsula, looks like the storefronts of an old western town, and is the site of the original post office and Lassen Drug, which dates from 1945. Some of the "older" is newer older. The creaky Timber House is from the 1960s, definitely from the motor lodge-motor court era. And before it was a grocery and gas stop for Lake Almanor visitors, Chester was a timber town, where the Collins Pine Company, which had owned forest land since 1902, began operating a sawmill in the

Tim Coonan

1940s. Lumber, cattle grazing, mining...the history of the West is one of commercial use of natural resources, with just a few exceptions (like national parks, a relatively recent development).

Although I now find patterns of land use and settlement interesting, none of this local history mattered to us at all, back in the day. Chester was relevant to us for two reasons. First, it had a Catholic church, and yes, our parents were the type who took their kids to Mass while on vacation. I don't remember grumbling too much about it, either. We were used to it, being an 8:00 a.m. Sunday Mass family ourselves. It was just something we did. And it was kind of interesting to go to Mass in a vacation town. We certainly didn't have to dress too nicely for Mass, that's for sure. I remember thinking, are these other families here also vacation families, or do they live here? If the latter, what do they think of us vacationers? My college girlfriend Mary, from Albuquerque, was also from a go-to-Mass-on-vacation family. Which she didn't mind. When they went to Mass in Laguna Beach, Mary observed that all the guys were in shorts (it was the OP—Ocean Pacific—era, after all) and were good-looking.

The second reason Chester mattered to us was actually tied to the first. After Mass Mom and Dad would take us for ice cream, literally sweetening the deal for going to Mass, somewhat compensating for the swimming time lost to kneeling-in-a-pew time. Lassen Drug had a soda fountain, that somewhat rare establishment at which not only was ice cream scooped but also incorporated into shakes and sodas. We had one in hometown El Segundo as well. The Gay 90s—obviously named before any judgment was attached to such a moniker—was an ice cream/soda fountain storefront on Main Street and was naturally a favorite of El Segundo kids. In any case, trips into Chester satisfied both our souls and our sweet tooth. But we still had to wait an hour before going swimming again.

Chapter 7

The Volcano on the Horizon

Although Lake Almanor isn't anywhere close to being a national park, there was—and is—a national park nearby, and Chester is kind of the gateway community for that, though not in any Gatlinburg sense. From Chester, as from much of Lake Almanor, Mount Lassen (or Lassen Peak) looms on the northwest horizon, and its importance to Chester is evident in the names of various businesses: Lassen Drug Store, the Mt. Lassen Club, Lassen Gift. Lassen is a national park. In fact, the park name, Lassen Volcanic National Park, recognizes the area's fiery, explosive natural history; no other national park is so named. Not only is Lassen a volcano, it's an active one, not extinct or even dormant. Now, I know this, because I teach geology and earth science to my middle schoolers, and I tell them, for one thing, that volcanism is the most exciting of the geologic disciplines. Sure, deposition is fine, and so is uplift and erosion (see the Grand Canyon), but nothing beats volcanism, and its parent, plate tectonics, for sheer geological excitement. Extinct volcanoes are those which have not erupted in recent history, and dormant ones show signs of activity but haven't erupted recently. Lassen last erupted in a series of eruptions from 1914 to 1921, making it an active volcano. And there are great black-and-white photos, even movie footage, of those eruptions, which apparently helped convince Congress to make the area a national park (Teddy Roosevelt had already established two national monuments in the area, by presidential decree).

Tim Coonan

Now, I was blissfully unaware of Lassen's active volcano status as a kid. Don't know if it would have bothered me. These days, we are all more aware of volcanic hot spots, thanks to the ubiquity of cell phone cameras and the ready access to videos on the internet. And I shamelessly tap into these resources when I teach volcanism to my middle school students. Who doesn't drop their mouth open in awe watching Mt. St. Helen's flank blow out, in an eruption which sent ash over most of the U.S.? I have shown the kids footage of lava from Mauna Loa, in Hawaii, slowly devouring a car and a mailbox. Does not disappoint. There is even a great animation, hour by hour, of Vesuvius raining down hellfire on Pompeii, forever changing those lives and that landscape. So now I am aware of volcanic hot spots in the U.S. and elsewhere, and it can give one pause. There are even massive "supervolcanoes," which last erupted hundreds of thousands of years ago, but which underlie popular recreation areas such as Mammoth Mountain in eastern California, and the Yellowstone basin (the latter being the basis for the disaster movie, 2012). God help us if either of those blows! Maybe Woody Harrelson will warn us.

Lassen Peak loomed large for us in 1975, and in 2016.

Lake Effect

All this is thanks to plate tectonics, the relatively recent observation that the earth's crust comprises fragments that are not at all stationary, but are in fact mobile, their movement driven by convection currents of hot, moving magma (magna is liquid rock; when it hits the surface, it's called lava). All the action occurs where the plates come together, crashing into each other like bumper cars, except that at most boundaries one plate is driven under another (oceanic crust being denser than continental crust). This causes friction, on a massive scale, which melts rock; under pressure, the magma, or molten rock, pushes upward and forms volcanoes. Mountains are also built at plate boundaries. In an extreme example, the India plate broke off from Australia, motored north, and slammed into Asia, pushing up the Himalayas.

Lassen Peak is not a solo volcanic performer here but is part of a huge volcanic symphonic arrangement. All around the edges of the Pacific, from South America to Alaska in the east to Indonesia and Japan in the west, plates relentlessly grind away at each other, their friction heat birthing a circumference of volcanos that encircles the Pacific Ocean. The aptly named Ring of Fire, a concept easily grasped by my middle schoolers (and anyone aware of Johnny Cash's music), is where 75% of Earth's volcanoes and 90% of its earthquake zones occur. Moreover, in the middle of the Pacific Ocean, the plate is passing over a volcanic "hotspot," creating, in succession, a chain of volcanic islands—the Hawaiian Islands. Pele and Maui at work. I have been fortunate enough to see lava spewing out of the ground at Mauna Loa, in Hawaii Volcanoes National Park—2,000-degree red molten rock spewing slowly out of a fissure, cooling quickly into those lava rock formations with great Hawaiian names: a'a and pahoehoe. It's one of the most spectacular sights you can see in a national park, and, as you might guess, it's tightly managed by the Park Service (because it IS 2,000 degrees). Even more spectacular are the spots where lava cascades out of cliff fissures and into the ocean, the cooler water immediately congealing the lava, turning it to rock and creating new land. As they say, Hawaii Volcanoes is the only national park continually expanding its boundaries.

Lassen Peak is not part of the Sierra Nevada, as you might think, but is actually in the southern extension of the Cascades, that volcano-birthed mountain range that dominates Oregon and Washington. In those states the range is marked by the volcanic giants of the Pacific Northwest: Mount Hood, Mount Adams, Mount Baker, Mount Rainier, Mount St. Helens. Lassen shares its southern Cascade neighborhood with its volcanic siblings, Mount Shasta and Medicine Lake Volcano; Lassen itself is the southern-most active volcano in the Cascades.

Though Lassen hasn't erupted recently, you can tell the area is volcanic in nature; there's stuff going on underneath the earth's surface there, a fact we discovered when my parents took us to Lassen on our initial Lake Almanor trip in 1971. It was our "activity" for one of our days up here; our parents were certainly prone to filling the week up with "activities," such as horseback riding and renting a patio boat (I suppose going into Chester for Mass and ice cream also counted as an activity).

The Bumpass Hell area is somewhat safer these days, thanks to a guardrail. But it still reeks.

Chapter 7

Rotten Eggs

That the place was different was apparent at our first stop, a spot called, fittingly, Sulphur Works, because it smelled like rotten eggs. This was something my dad, a chemical engineer for Standard Oil, was very familiar with; one of the plants at the refinery was solely focused on removing sulfur from fuel oil. Sulphur Works is a hydrothermal feature where the innards of the volcanic earth bubble to the surface. Early settlers mined sulfur here, and the area is actually part of what's left of a very large volcano, Mount Tehama, which erupted almost 400,000 years ago, and is now just an eroded crater. We just knew that it stank.

Further up the road we stopped at Bumpass Hell, a feature whose name we, being good Catholic kids, absolutely delighted in. The names of Lassen features are wonderfully descriptive: Brokeoff Mountain, Chaos Crags and Chaos Jumbles, Devastated Area (what more do you need to know?). Red Cinder Cone, Bathtub Lake, Boiling Springs Lake, Devils Kitchen. And Fantastic Lava Beds. Who wouldn't want to see that? Bumpass Hell, though...obviously named for its supposedly hellish landscape. Of course, many hydrothermal features are actually quite beautiful. Mammoth Hot Springs in Yellowstone is a towering, cascading feature where bubbling hot water continually deposits minerals, building up the deposits. And Grand Prismatic Spring is very aptly named, just a kaleidoscope of colors. As for Bumpass Hell...sure, the actual feature is barren, because what self-respecting plant would grow on that mineral desert? But it's in the middle of mountain meadows with conifer-framed views of the volcanic remnant mountains.

Tim Coonan

Bumpass Hell contains fumaroles and mudpots and boiling springs, and here the volcano beneath sends up volcanic gases which bubble and hiss and stink, reminding one, somewhat gently, of the sleeping giant beneath. Don't fuck with me. Not unlike the snoring of the dragon Smaug, perhaps.

A rare photo: the only hike we ever did a as family, to Ridge Lakes at Lassen.

Further up the road (though still not very far into the park) we got more ambitious: we went on a family hike. Now, just as we were not a boating family, we were also not really a hiking family (or a camping family), by any stretch of the imagination. No huge Coleman tent or two-burner Coleman stove in our garage. But we boys were in Boy Scouts and had gone on hikes and campouts, clad in Scout green with our carefully-rolled Scout neckerchiefs and outfitted with the very non-technical (compared with current REI-sourced) gear of the 1970s. Terry and I had metal canteens, Army-issue (one of ours was my dad's from his Army Reserve days), which hung on broad web belts, and I can still taste the water from that metal canteen. And now we've passed all the way through the plastic/Nalgene water bottle phase back to metal water bottles such as Hydroflasks, and I love it. Tastes good. We Boy Scouts had compasses and matches in waterproof containers; snake bite kits (never used) with suction cups to draw the snake poison out, and bands to constrict blood flow. We knew how to apply tourniquets (again, never used). We had signal mirrors and space blankets in case we got lost. My friend Mark had a fire-starting kit with flint and tinder, and became pretty good at starting a fire that way. We hiked with bulky, non-adjustable, canvas metal frame packs (Kelty became the gold standard) and slept in tube tents, really just a tube of plastic sheeting hung on a rope between trees. And there were NO pads, inflatable or otherwise, to cushion our inadequate sleeping bags from the hard ground. But we were young then, what did we care?

It was "only a mile" to Ridge Lakes, two small lakes on the saddle between Mount Diller and Brokeoff Mountain. Though the trail was only a mile, it was a thousand feet of elevation climb up to the lakes, and current hiking apps call the hike "strenuous" with "no level points" (one hiker's comments: "This trail will kick your ass"). Suffice to say we didn't know what we were getting into. At the time, with little experience in topography, we thought the trail must have been misnamed; they must have meant it was a mile in elevation gain (it certainly seemed to be). As I recall, we stopped some part of the way up, and took a family photo on a log, Then Dad and the boys went on to the lakes—which were beautiful, high-elevation mountain lakes; Mom and the girls, as well as baby Dennis, stayed behind on the log.

By 1994, the photos were in color, and the volcanic lake was as yellow as Easter egg dye.

Tim Coonan

I remember being amazed that there was snow, in midsummer, by the side of the trail. And I suspect we didn't bring much, if any, water; Terry recalls that the lakes were too acidic to drink from (lakes near a volcano can become acidic from volcanic gases, such as carbon dioxide and sulfur dioxide, going into solution in the lake waters). According to Terry, Dad treated us to Orange Crushes when we got back from the trail. Terry, who admits that many of his vacation memories are about food, also says we ate dinner at a restaurant in Chester that night, where, for the first time, we had Cornish game hens. Which were very good, apparently. To my memory, we never again went on a family hike.

But some years later, I found myself hiking to Ridge Lakes again. In 1994 Cathy and I met her sister and family near Lassen (and Almanor!) for a few days. And the color photos sure tell the tale: the Ridge Lakes were acidic! Yellow-greenish water, obviously full of God-knows-what chemicals. No, I wouldn't be tempted to drink from those lakes, which were fine for wading in and skipping stones.

Lassen's base is anchored by reflective Lake Helen, and snow never melts from its ridgetop.

And many years later, my daughter Bridget and I hiked to the peak of Mt. Lassen itself. Needless to say, we took plenty of water, and food, and were well aware of the distance involved, and the elevation gain. As far as hikes go, it wasn't

Lake Effect

THAT arduous; five miles round trip, but with 2,000 feet of elevation gain to the peak, which tops out at 10,457 feet. The National Park Service calls the hike "strenuous," and I suppose it was, or at least more so for me; Bridget's lungs, legs and heart are 35 years younger than mine! From the trailhead at beautiful, blue, glacial Lake Helen, the trail winds and switchbacks up a series of ridges, overlooking the lake, then climbs above tree-line, past the last of the few hemlocks and pines, to the rocky, scree-strewn slopes of the mountain's shoulders. The last mile or so the trail crosses large patches of snow (at least that summer) and winds between large boulders to the summit. And butterflies. Every summer, apparently, millions of bright orange tortoiseshell butterflies make for the upper slopes of Mount Lassen and other peaks, maybe to escape the valley's heat, maybe for breeding, their larvae feeding on certain species of *Ceanothus* (buckbrush). It happens every year, not unlike the mythical return of the swallows to Mission San Juan Capistrano. But there they were, thousands of them swirling around as we crossed the snow patches up high, their orange bodies standing out against the white snow. I did not expect that.

Bridget and I enjoyed the high from the summit, and the forgiving view of the world below.

Tim Coonan

Lassen Peak is a lava dome, a hardened mound of lava that has pushed up its crater. In fact, it's the largest lava dome in the world, by volume. The view from the top is commanding, and includes the nearby lesser peaks of the park, but also its sister volcano, Mt. Shasta, to the northwest. There, is something immensely satisfying about summiting a peak. There's the accomplishment, of course, and the view, and the relative rarity of it. For millennia we humans have wanted to climb that ridge, or that mountain, to see what we can see from the top. And everything looks pretty good from on high. The Earth is still there, spinning beneath us, living and breathing and just being, despite all the crap we've done to it and each other. Bridget and I took in the view for quite some time. Who knows when you might be back this way, after all?

Chapter 9

Whiskey's for Drinking

And water's for fighting, as the saying goes, and that pretty much sums up the history of water in the West. Almanor is no exception. Almanor wasn't always a lake—and I realize you can say that about ALL lakes, given plate tectonics and hundreds of millions of years. Indeed, there are places that WERE lakes once, but aren't anymore, such as Death Valley: currently a playa, a dry lakebed, Death Valley was once the site of Pleistocene Lake Manly. But in the case of Almanor, we aren't going back as far as the Pleistocene. Almanor hasn't been a lake very long at all. Barely over a hundred years; a dam created Lake Almanor in 1914. And the story behind the lake's creation is one that surprisingly (or not, depending on your level of cynicism) involves subterfuge, backstabbing, and maybe even arson, in a plot so twisted that only Jack Nicholson's Jake Gittes, of *Chinatown*, could untangle it.

Now, many, if not most, of the larger lakes in California are human-made, with a few notable exceptions, such as Tahoe, that behemoth lake which dominates the "elbow" of the California-Nevada border. Tahoe is a graben lake, in which the basin is formed by a combination of uplift and subsidence. There are also glacial lakes, formed by the slow but powerful grinding of glaciers over terrain, and the Sierra Nevada abounds with them. Finally, there are volcanic lakes, such as those found near Lassen, formed by the damming effects of lava flows, or by water filling the collapsed crater (the caldera) of a volcano. The best example of this is Crater

Tim Coonan

Lake in Oregon, which happens to be the deepest lake in the United States.

But Almanor is a reservoir, a human artifact, like Shasta Lake and Lake Oroville, and those lakes along the Colorado River, such as Lake Havasu. Now I knew, early on, that Lake Almanor was drowned land. When you boat or paddle there, you can sometimes see the dead trees, the snags, eerily passing by in the clear lake water below. What I didn't realize was that most of what is now Lake Almanor was grassy, verdant meadow, a meadow so large that the early settlers called it, appropriately enough, Big Meadows.

Of course, early settlers weren't the first folks to use the area. In fact, they weren't even the first settlers. When you see the word "settlers," you think of sturdy, stoic white folk, stern womenfolk in bonnets and gingham dresses, and riding buckboard wagons. Why can't that term also include the sturdy brown folk who lived on the land long before Anglo "settlers" arrived? In the Lake Almanor region, these were the Maidu, who were indeed settlers; they were not nomadic, but lived in scattered villages, each of which comprised a small number of their branch and bark shelters (hubos). There may have been ten or so of these villages ringing the big meadow, set in the shelter of the trees. The Maidu were known for their beautiful basketry (I imagine the meadow was a great source for reeds) and were part of a larger native grouping which we have derogatively called "digger Indians" for their use of wooden digging sticks.

It all comes down to technology, doesn't it? Guns, germs and steel, as Jared Diamond says. The Maidu way of life was gradually erased from the Big Meadows area, no match for the technology and organization and zeal of the Anglo onslaught (and likely some germs, too). First contact was probably with emigrants on the 1848 Lassen Trail, basically a shortcut from the Oregon Trail for those who were cutting through the area on the way to the central valley of California. Stumbling upon Big Meadows, which the Maidu

Lake Effect

called Na'kam Koyo, it must have seemed to them a little slice of heaven: bubbling springs, grass for their stock, the North Fork of the Feather River running through it. Plenty of fish and wildlife for sustenance.

The California gold rush in the 1850s pushed would-be miners up the North Fork of the Feather River toward Butt Valley, Humbug Valley and Big Meadows, with the first Anglo settler putting roots down in Big Meadows in 1855. Within a decade the Maidu in Big Meadows were working as ranch hands for the new ranchers, though they still made beautiful baskets, trading them to Anglo settlers and tourists. But the die was cast. It wasn't exactly the Oklahoma land rush (which also came at the expense of Natives), but the Big Meadows area was quite thoroughly Anglo-settled by the 1870s. An 1892 map of Plumas County shows the Big Meadows area somewhat thickly occupied with homesteads, which included dairy farms (some run by hardy Italian-Swiss immigrants) but also hotels; by the 1880s tourism was thriving in the area, as folks traveled here by wagon for views of Lassen Peak and for pack trips into the Lassen area to see the volcanic features. The town of Prattville, on what would later be Almanor's western shore, was well-established by then, and was the site of several hotels. The map also shows, interestingly enough, that much of the timbered area away from the meadows was now owned by the Sierra Lumber Company, which is to be expected; it was only a matter of time before the timber companies expanded their operations into the area, aided by the extension of railroad line. Dairy would yield to timber as the primary way that economic gain would be wrung from the area's natural resources.

Grazing, mining, timber extraction...is there any part of the West that hasn't been impacted by these landscape-altering uses? Actually, there is; we call them national parks. The Big Meadows area was not immune from such uses, from the drive to extract riches from the land in ways the native Maidu never dreamed of. And then there was water. Maybe the ultimate natural resource in the arid West, it made all other natural resource use possible. Water to support grazing. Water to drive sawmills and log flumes. Water in which to pan for gold. And water to supply power, to produce that newfangled electricity, destined soon to power a nation.

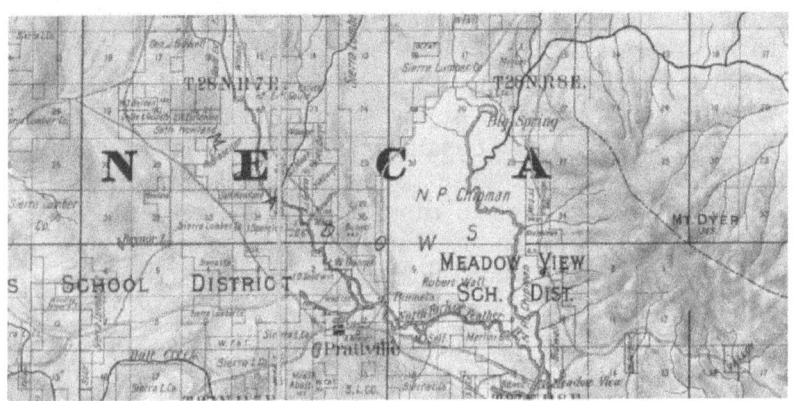

A pre-lake map from 1892 shows both the North Fork of the Feather River and Big Springs, both to be inundated (as would be the town of Prattville).

Big Meadows had plenty of water. The North Fork of the Feather River flowed through it, rising to a torrent in the southern end of the meadow, where the confluence of the western and eastern branches met, the latter fed by Big Spring at its upper end. From the confluence the North Fork hurried downhill to meet the Feather River in the heavily wooded canyon to the south. But I hazard a guess that the area's potential for an electricity-producing dam was considered by none of the Big Meadow residents, and by very few other people at all. Except for one.

Julius Howells had traveled through the Feather River area, including Big Meadows, in 1882, as a young geology

Lake Effect

grad student, on an expedition led by Harvard geologist Alexander Agassiz (son of noteworthy biologist Louis Agassiz) to study the geology of the Feather River area. Howells must have been impressed by the Big Meadows area. He returned to the area as a civil engineer in 1901, having studied hydraulic gravity-fill dams in the interim. At some point it had dawned on him: the Big Meadows area and the North Fork of the Feather River had enormous potential for producing electrical power, through construction of a dam. The Feather River was the largest river issuing forth from the Sierra Nevada-southern Cascades, and the North Fork was the largest of its tributaries, with a significant drop in elevation from Big Meadows to Oroville. As I teach my students, things at a great height have great potential energy, which can be converted to kinetic energy when they fall. Howells saw that on the North Fork. It was his idea to dam the North Fork at the southern end of Big Meadows, producing significant hydropower and creating a large reservoir.

Of course, Howells was just a civil engineer; he needed backers and big money to pull this off. He found them both in the Earl brothers. Wealthy and influential, Guy Earl was an attorney in Oakland with connections, his brother Edwin a wealthy citrus mogul in LA. It didn't take much convincing. Introduced to Howells in 1901, by 1902 the Earls had created the Western Power Company and had hatched a plan to buy up the Big Meadows homesteads and ranches, surreptitiously, so folks wouldn't be tipped off and raise their selling price. This is a pretty standard MO for land grabs. Think of Walt Disney secretively buying up orange groves in Anaheim, later to become Disneyland, or John D. Rockefeller buying land near Jackson, Wyoming to create Grand Tetons National Park.

To pull this kind of thing off you need a mole, a man on the inside, and Howells and the Earls had found one willing to work for them. Bidwell was a name that had been associated with Big Meadows since 1860, and, it seems to me that

Tim Coonan

almost all members of that family had a gleam in their eye, had figured out ways to milk money from the land. In the 1860s Bidwells were buying up land in Big Meadows, and had built a large hotel (Meadow View) at the southern end, where they were collecting tolls on the nearby road and bridge that crossed the North Fork. Bidwells invested in mining operations and foundries, and were the driving force behind a dam at Round Valley Reservoir, used to support mining operations, and where Bidwells owned the water company. Water, tourism, road building, mercantile, mining, even telegraph operations...the only thing missing from the Bidwell portfolio seemed to be timber. So it's little wonder that Gus Bidwell, who by 1900 was the successor to Bidwell Big Meadows interests, smelled money when the Earls and Western Power Company came calling. Gus Bidwell became not only the first landowner to sell to the power company, but also their confidante. He convinced other landowners to sell to the shadow company buying the land, ostensibly for a large ranch, like the King Ranch in Texas. By 1902 Bidwell and the Earls had obtained buying options on over 11,000 acres of Big Meadows ranches. For these efforts Bidwell was rewarded; he became director of operations for Great Western Power's interests. It came at a price. Bidwell would also later be viewed by locals as a turncoat, his legacy now cemented as a dishonest facilitator of Western Power Company surreptitious land grabbing. What a tangled web we weave.

It gets better. To build a dam and use the water, the company needed to post claims for water rights, as per the law at the time. And this became a footrace (well, a wagon race) as it turned that, in fact, someone else HAD thought of the same thing, that Big Meadows had enormous potential for hydroelectric power. Howells met Bidwell at nearby Greenville in April 1902, and the two set off by wagon for Big Meadows to post water rights. They were in a bit of a hurry; Howells had observed two men on the train from San Francisco who he

Lake Effect

suspected were surveyors, based on their footwear (high-topped boots commonly worn by engineers and surveyors). Howells and Bidwell nailed up signs in southern Big Meadows, near the future site of the dam, laying claim to 2,500 cubic feet per second of North Fork water for various purposes, including hydroelectric power. Downstream they found two men posting their own claims for water rights; the two were working for G.P. Cornell of Greenville, who in fact had his own plans for hydropower in the area. Although Bidwell and Howells technically had the rights, having already posted them, it now became a race to record the rights at the county seat in Quincy. Some mayhem ensued. Howells and Bidwell tore down the area's telegraph line to prevent Cornell's men from telegraphing their claims in. While Howells stayed behind to post more water rights, Bidwell rushed on to Quincy, rousting the county recorder from his home at 8:30 p.m. to go to the county office and record the claims. Cornell's men arrived at 9:15, having lost the race by about 45 minutes. Great Western Power now held the water rights to Big Meadows, thanks, in part, to a smashed telegraph line.

Great Western Power, the "Great" having been added to the name after significant backing came from New York City financiers, bought Meadow View and the Bunnell and Prattville hotels. Their plans had expanded: the size of the dam and the reservoirs increased, as big money smelled more. The larger planned reservoir size now required Great Western Power to acquire all of Prattville; instead of becoming prime lakeside property, Prattville would now be underwater. Many in Prattville did not want to sell, and here's where it gets murky, maybe even nefarious. Jake Gittes stuff. On July 3, 1909, most of the town residents were attending a Fourth of July celebration in a meadow outside of town, when black smoke appeared above the trees. The town was on fire! Pretty much the whole town was subsequently destroyed, consumed by the blaze; firefighting ability was

Tim Coonan

lacking, given wooden buildings, the lack of firefighting equipment and poor water delivery capability (and years later, local towns would still be vulnerable to fire, even though those capabilities had increased). The cause of the Prattville fire was never discovered—maybe today's arson investigators would have been able to pinpoint a cause—but it was widely suspected that Great Western Power was behind the blaze. Never proven. But the company sure benefited from the fire: Prattville, never rebuilt, became a ghost town, and Great Western Power was able to force land sales though condemnations. You can't fight city hall. Or, apparently, Great Western Power.

With all obstacles and reluctant owners overcome, Great Western began dam construction in 1910. The original plans called for a concrete arch dam, and five of the arches were actually constructed, when it was determined that there was inadequate bedrock to support such a dam. Seems like an inexact science, doesn't it? Holding back all that water with a design that couldn't be trusted? Today we trust dams to hold, and I have heard of nobody nervous about Hoover Dam or Glen Canyon Dam failing. Great Western Power was not to be deterred by the lack of bedrock. Howells quickly designed an alternative, shifting from a concrete dam to an earthen fill dam. Which, to me, sounds even less trustworthy. So let me get this straight: you're going to dump tons of rock and gravel in the river bed, inundate it with wet clay soil, and use it to hold back millions of gallons of water, all of which desperately wants to flow downhill? Hmm. But these dams of the early 20th century are engineering marvels. Built long before computers and laser sighting tools, they have successfully held back billions of tons of water and produced countless megawatts of electricity, with no hint of failure (okay, maybe one: in 2017, a spillway failed on the earthen fill dam at Oroville, downriver from Lake Almanor, causing the evacuation of close to 200,00 area residents).

I'm not saying dams are a good thing. They have disrupt-

Lake Effect

ed native fisheries and ecosystems and destroyed—forever—beautiful canyons, even inundating archeological and historic sites. They've provided power that we take for granted, and that has allowed the monstrosity that is Las Vegas, Edward Abbey's "Glitter Gulch," to grow unfettered. And, of course, these dams and reservoirs are built on at least one failing principal: that precipitation will be sufficient to keep their reservoirs filled. That has recently been given the lie, as we see "bathtub rings" appear, and deepen yearly, around all the major reservoirs of the West. Lake Mead, in fact, will never be full again, victim of a 20-years-and-counting drought gripping the Southwest. That drought could even become one of the "megadroughts" that periodically appear in the Southwest, often for 50 years or more. One such megadrought in the 13th century may have contributed to the demise and disappearance of the pueblo-dwelling Anasazi. Not that any drought will make Las Vegas disappear.

The new dam site at Big Meadows was just upstream from the original, and construction on the new dam began in 1912. When finished, it would be the largest earthen fill dam in the world. It wasn't quite ready to be filled in 1914, but Mother Nature had other ideas; nature bats last, as the old Earth First bumper sticker said. Turns out the winter of 1913–1914 was one of heavy precipitation. Rain and melting snow in spring 1914 began rapidly filling the reservoir, catching Great Western Power a bit unprepared. The company's MO had been to burn ranch buildings on properties it had bought, as a cheap way to clear the soon-to-be-flooded land, but the rapidly rising reservoir waters prevented them from doing this in many cases. Buildings were left standing in the rising water, and there was even a report of 100 cattle being stranded on a hillock ("they might have to swim for it," said a local newspaper account). So be careful what you ask for—you wanted a lake? You got it. There was no stopping the rising waters.

Tim Coonan

Then there was the issue of lumber. Although it was indeed a meadow, there was plenty of timber in Big Meadows, by some accounts as much as 80,000,000 board-feet in the area to be inundated. The newly created Red River Lumber Company, which had built a sawmill in Westwood, north and east of Big Meadows, had contracted with Great Western for the timber rights to the lake area. Crews began furiously felling timber in the Big Meadows area, in a race to salvage as much as possible. Felling crews left the down trees in place. Rising lake waters would float the logs to the surface, and for some years later barges could be seen towing floating logs to Hamilton Branch at the upper end of the lake, for rail transport to the mill at Westwood. No reason to leave all that money at the bottom of the lake.

The dam was completed in June 1914, but that wasn't the end of the story; it grew higher over time, and the lake expanded. The dam was raised in 1916, and an entirely new dam was constructed in the 1920s, just downstream from the original. The lake grew from its original 220,000 acre-feet to its current 1.3 million acre-feet because, well, why not? Money follows money, and several corporations and a significant number of people made a lot of money off of Lake Almanor. Oh, the name...Julius Howells, in an all-star piece of brown-nosing, suggested naming the new lake after Guy Earl's three daughters, Alice, Martha and Elinor. Imagine having a lake named after you. Might as well name it Lake Hubris. But I can't be too disparaging of this. I've thoroughly enjoyed the lake they created.

Building a dam to stop a river in its tracks is one of the most brazen and audacious acts ever conceived by humans, a veritable thumbing of the nose at nature. Sure, beavers do it, and quite well, but on a much smaller scale (and sometime beaver dams really piss off the humans affected by them). There is a possibility that Nature itself noticed—and reacted to—the new dam and lake in Big Meadows. It's thought by some that the sudden appearance of the lake and

Lake Effect

its billions of gallons of water affected the local geology, the volcanic nature of the region, resulting in the 1914–1917 eruptions at nearby Mount Lassen. Perhaps the volcanic giant was just saying welcome to the neighborhood, but I'm still the law in these parts.

Chapter 10

The Gold Standard. Or Shangri-la

As my mom lay dying in 2020, midst-pandemic, all of us six kids were home, together for the first time in a while. Dennis and I found ourselves going through and scanning old photographs, many from my mom and dad's respective childhoods, which had never made it into an album or seen the light of day for many years. We also found a DVD that Dad had made. Some years prior—and he didn't remember doing this—he had had all our home movies, shot with 8mm and Super 8mm film, transferred onto a DVD. He said he had to take them somewhere in Santa Monica to do this. The movies were clipped together into segments, with basic titles, and, since there was no sound, the footage was accompanied by instrumental music. Much of it Irish. And much of it emotional. Yes, "Danny Boy," which *Saturday Night Live* famously dubbed the "Irish crying song," was one of them. As if watching that old footage wasn't emotional enough. I resolved to make copies of that DVD for my siblings.

However, that DVD, much like the One Ring, was lost, passed out of time and memory. I spent time looking for it when I visited my dad, but to no avail; I thought the DVD, and the footage, was lost forever. The original 8mm and Super 8mm reels, the prime sources, were long gone, having been purged, likely along with the old 8mm projector, which I remember Dad carefully threading the delicate film into when we watched family movies. That was a dicey operation, which could go wrong at any moment, with the specter of destruction palpable, hanging in the air. A lot was riding on

Lake Effect

that operation, which I learned to do myself.

A year later Dennis found the DVD. Hiding in plain sight. I wasted no time uploading the files to the Cloud, in an attempt to make them live forever, and I was able to get my dad access to them on his phone. The first video is footage from my parents' wedding in 1957, shot at the reception (no way Monsignor Giusti would have allowed any such thing at the wedding Mass itself) with subjects illuminated by a bank of lights just slightly less bright than the outfield lights at Dodger Stadium. But there's my grandfather, looking dapper in a white tux jacket and black bowtie, with that twinkle in his eye. My grandmother in a beautiful blue flower dress. Uncle Ted and Aunt Carol making each other laugh. And my parents—so young!—bright-eyed and hopeful, with decades of family happiness ahead of them. My dad watches that video several times a day now, and he says it's great to see Mom so young and alive.

There's also Almanor footage on there, my dad having faithfully chronicled, in both stills and movies, all the important family activities in what I call the Golden Years, from their wedding to our college graduations. Babies, Little League parades, flag football games, family vacations. The Almanor footage from 1975, our second trip to the lake, was pleasantly startling to see, because there it was! The Hillyer house! With its unmistakable squared-off A-frame design. Look, I know it was just a cabin, a house on a lake. Not huge, not well-appointed. Funky, even, with that A-frame design. I don't even know if it had air conditioning. No huge deck. But sometimes a place becomes bigger than life, certainly bigger and more important than its size or design warrants. Our childhood homes can become like that. I remember our backyard as being big, big enough to play football and baseball in, even with a peach tree and large swing set occupying portions of it. It's not big. When I visit my dad these days and look at that backyard, I wonder how we didn't hit every ball over the fence.

Tim Coonan

The Hillyer house in its heyday, 1975.

The Hillyer house became such a place, the one conjured in our minds when we remember Almanor. Our minds don't go to the Knotty Pine Lodge of the previous trip, or the nondescript cabin we rented in 1978; I can't remember that place at all, really. But the Hillyer place...what made that place, or that experience, so idyllic? I can't really say. Certainly, there's the setting, for starters. It was located near the tip of the peninsula, previously the ridgeline that bifurcated Big Meadows, and the further you go down the peninsula, the less traffic there is. The tip of the peninsula itself is a great place to walk to, where you can enjoy a pretty-much 270-degree view of the lake, from views of Lassen Peak in the northwest to Dyer Mountain and Hamilton Branch to the east. And at the Hillyer house, we were allowed to walk down to the peninsula tip by ourselves. The houses at the end of the peninsula are away from the hustle and bustle of Big Cove, where boats go in and out from dawn (the fishermen)

Lake Effect

until after dusk, and where you had to walk down to the public docks to go swimming. But at the Hillyer house we had our own dock! A simple wooden floating dock connected to the rocky, sandy shore by floating walkways, it became the center of activities for us. It seems as though every minute of the week, at least one of us was out there swimming or just enjoying the dock. From the modest cabin a grassy lawn sloped gently down to the shore, and the dock, and in video footage there's 16-year-old Terry, heading out the dock, trailed by 5-year-old Dennis. Terry's back is broad from weightlifting; about to enter junior year of high school, when he'd be a starting guard for the football team, ascending to team captain his senior year. A beach towel is draped over his shoulders and he carries an air mattress; we all had them: thin rubber things that bent readily in the middle and punctured easily. In other footage Terry is on an air mattress in the water, and is wearing a swim mask, a black round one as worn by Lloyd Bridges in *Sea Hunt* or commonly seen on *Flipper*, one of our favorite shows (do those shows define an era, or what? *The Rifleman, Branded, Bonanza, High Chaparral. F Troop, Hogan's Heroes, The Wild, Wild West...* how did we get anything done back then?). Terry slips easily off the mattress and dives into the lake. To see God knows what. Maybe tiny minnows; lakes are not known as great snorkeling spots. It's not exactly a coral reef down there. In the video Dennis is pursuing his own simple lake pleasures —he's carrying a good-sized rock to drop in the water.

The dock. Believe me, we'll return to this subject, and I will probably refer to it then as the Dock, capitalization according it the respect and honor it deserves as the center of future Almanor trips for us, the Coonan kids, the next generation. Who would use the Dock as the centerpiece of Almanor weeks, not for the traditional dock uses, such as boat mooring and fishing, but as the ultimate family gathering spot. A proverbial big tent, family room, kiva...whatever communal family gathering spot comes to mind for you. Dare

I say church? Maybe. And that begs the question: what gods are worshiped on the Dock? Some of you know the answer to this: Springsteen, Buffett, the Eagles, to name a few of the major deities.

First known image of Coonans on the dock.

The pic of the dock in 1975 is pretty Norman Rockwellesque. It is the first known picture of Coonans on the dock, and it's fitting that the still from the video footage is fuzzy, low-resolution, blue-toned, because use of the dock would evolve from there to a state of high sophistication and ritual. But in 1975 I was just discovering the pleasures of the dock, and they were distinct and simple pleasures, which I still recall to this day. Look, life is stressful enough, and we all have happy spots we return to in our minds, spots that maybe offer instant relaxation. The dock at the Hillyer house is one of those for me. Fifty years later, I can still remember lying on that dock, in the summer sun, looking up at the tall pines framing the view of the sky (unsullied by wildfire smoke). The dock moves up and down rhythmically, pumped gently by the very small lake waves on their way to shore. I think I first realized the pleasures of just sitting in the sun

Lake Effect

then. I've no doubt done a bit too much of that over my life, given my Irish skin; my Celtic ancestors needed every drop of sun-produced Vitamin D they could get in their northern clime, and have no protection against the sun's rays. You plop that Irish skin elsewhere and it just burns, and years later you have that skin damage that your dermatologist scolds you for. Didn't help that I grew up a mile from the beach in southern California (our idiotic motto: you have to burn before you can tan), and chose a profession (wildlife biology) that took me outside. But in fact, that's one of the reasons I chose that profession! Not built for an office, am I. And I have no regrets.

Now, my mom had alabaster skin which never picked up color, because, for one, she never spent much time in the sun. Or was covered up when she did, smart woman. So much so that a family doctor once told her she needed more sun. She certainly hectored her sun-loving kids about wearing sunscreen. But there she is in the Almanor footage, floating precariously on an air mattress near the dock, in that same pink one-piece and, yes, wearing a bathing cap on her head. She smiles and squints into the sun. In another scene Dad helps Dennis lie on an air mattress—one of the few family movies in which Dad appeared. Maybe I filmed that.

Besides sun-worshipping, also filed under not-so-bright things I have done, is a wildlife encounter I had at the Hillyer house. On the shore, near the dock, I saw a small bat struggling along in the very shallow water, among the small rocks. Had it fallen in the water? It wasn't flying, but was walking on the shore during broad daylight. Should've tipped me off that maybe the bat wasn't all right. I know now that bats can carry rabies, and at a higher rate than other mammals, and can even transmit diseases such as histoplasmosis. These facts are commonly known and have been used by otherwise well-meaning folks, even National Park Service staff, to cast aspersions on bats and their roosting presence in NPS buildings (their natural roosting sites, such as tree cavities in large trees, being greatly reduced in number. I have had to defend the presence of Townsend's big-eared bats in historic buildings at Channel Islands National Park). But in 1975 I, future wildlife biologist, was just fascinated by this bat; I had never been this close to one before. I went back to the cabin and Mom gave me a coffee can. I returned to the shore and put the bat in the can, risking rabies and Ebola and maybe even Covid-19 in the process, and I can't tell you what happened after that. I probably came to my senses—or Mom did—and I ended up leaving the can and the bat out on the rocks, where either nature (red in tooth and claw) took care of the situation, or the bat hobbled/flew off.

Other Almanor wildlife were documented by Dad with his movie camera. There is footage of young mule deer bucks, their antlers in summer velvet, padding along the shore and in the yard at dusk. We loved it. Dad even caught a bald eagle in flight over the lake, a pretty good trick given the poor zoom capabilities of the camera and the eagle in flight. To this day you can see bald eagles at Lake Almanor. A pair nests in tall pines on the peninsula, near the end, and local residents such as our friends Joe and Michele O'Neil know where they nest each year. This makes it a homecoming of sorts; you can count on the eagle pair nesting at Lake Almanor.

Dad took considerable footage of our patio boat lake cruise that summer. Makes me wonder how many rolls of film he used—I can see that camera now, metallic silver, with a large key that you turned to open the case and replace the film; the key would then flip flat and lie recessed in the camera side. The whir of it filming, and, weeks later, the whir of home movies being projected in the living room. Of course, no sound was recorded, so all the scenes are inherently peaceful, and no subject wastes any breath talking to the camera. Kinda nice, when you think about it. So there's Lassen Peak across the lake, a little fuzzy (we are so used to HD, aren't we?) in a light blue sky—not a hint of smoke—and with plenty of snow on its slopes. Very few boats on the lake. And then there are the required shots of each one of us driving the boat. Katie in pigtails. Dennis on Dad's lap. Dan in his El Segundo Little League All-Star cap. He missed part of his all-star season to be on this family lake vacation, which I'm sure vexed him—and no doubt his coaches—to no end. While Katie is driving the boat, Terry is deep in a book on the deck, voracious reader that he was, and is. He sits in a cheap, webbed, aluminum lawn chair—those were the only seats on the boat! Patio boats have come a long way.

And then there's me. Sun worshiper, my shirt is off as I sit behind the wheel of the boat—in a lawn chair—and I'm considerably skinnier than football-playing Terry. I still hadn't figured out that I had a body much more suited to track and cross-county, though I'd eventually get it. Terry and I, though twins (fraternal), had very different body types. If we had been our peasant Irish ancestors, Terry would have stayed put and fought the British usurpers; I would have run the fuck away, into the hills. In the footage you can see the thin, wispy mustache I had grown that summer. Hey, it was the 70s, after all. My friend Rick, upon seeing it, asked me, "Where'd you get that mustache? From sucking armpits?" No, it was not very impressive, and neither was it long-lasting. Though it survived until at least picture day at high school. And if my mom were around, I'd ask her now what she really thought of it.

Dad also took some footage inside the Hillyer house, because it had an unusual feature: a firepole descended from an upstairs bedroom into the living room, delivering an erstwhile firefighter to a pile of pillows on the floor. Dennis, young climbing daredevil, loved it (we all did, actually). In the footage, Dennis slides down the firepole, in his pajamas—three times! Apparently, Mr. Hillyer was a fireman at the China Lake Naval Weapons Center, near Ridgecrest, and a firepole made it up to their vacation house at Lake Almanor.

Iconic Christmas card image from 1975.

Dad had us six kids out on the deck for another scene. We lined up and sat on the deck railing, Dan, again, in his All-Star hat, Katie in pigtails, me with 70's hair, OP shirt and my Onitsuka Tigers. Terry in Jack Purcells. Dennis is sucking his thumb and Nora is cutting up. This photo, which ended up in a Christmas card that year, became the iconic photo of us at Almanor, perhaps the one we remember most when we think of the lake. You can see the A-frame shape of the cabin, the smaller upper deck where we hung our towels to dry, lawn chairs on the deck. And this would be the last time all six of us kids were at the lake until almost 40 years later—Terry didn't make it to the next Almanor trip, in 1978, because he had a summer job at Penney's. Sometimes you don't realize the significance, the gravitas, of a family moment, until it has passed. But then you remember it fondly.

Chapter 11

The Return of the King

Dan, in particular, remembered Almanor fondly. For years, ever since he was a kid. Now, the rest of us did, too, but I think Almanor stuck in Dan's mind, like a song fragment or a bad thought, as something that he just could not get rid of, something that was always on the back burner. Dan says that, like me, he has an inordinate respect for anything we went through as kids, and he was always infatuated with the idea of Almanor. Was it as cool as we thought it was, as kids? Or was that just nostalgia, viewing the past through rose-colored glasses?

Dan moved back west in the mid-1990s, to the Bay area, where he worked (in athletic administration) at Cal, at the then-Pac-10, and ultimately as athletic director at Santa Clara University. For folks in the Bay area, Tahoe is the big lake draw, and Dan met many who vacationed there on a regular basis. Dan and Donna even spent some time there themselves, with their growing family. There was a Cal donor whose Tahoe house they stayed at, and another family in San Jose who had a house up there. As beautiful as those houses were, and as the spectacular Tahoe setting is, Dan was never sold on it. Because, for one, you can't swim in Tahoe. It's just too damned cold. It's one of the deepest lakes in the U.S., and it just never warms up. I can attest to that. Went there with my high school friends Mark and Vito one summer, during college, and I couldn't even walk all the way into the lake. As Vito described it, it was the phenomenon of "crunge," which he claimed was a Led Zeppelin term for when your testicles, having been briefly exposed to extreme cold, crawl—no, flee—up into your body cavity in a desperate act of self-preservation. Call it what you may. But it certainly happens in Tahoe water and is no doubt somewhat related to the Seinfeld phenomenon of "shrinkage."

So, no swimming at Tahoe, even for kids, who are prone to go swimming no matter how cold the water is. Furthermore, Tahoe is crowded, boats and people everywhere, to the point where you can't get a dinner reservation, and then have to wait an hour and a half for dinner. That's a non-starter for a family with young kids. And the rentals are expensive. Then there's the traffic: you're fighting Bay area traffic to get there on a Friday afternoon, and again on a Sunday afternoon return trip home.

The years passed, and Dan thought of Lake Almanor more and more. By this time, at the San Jose parish school their kids attended, they had met several families who were tied to Lake Almanor, even had family homes there. Trish Foster had massive Christmas parties every year, not unlike Nora's, and the O'Neils—Almanor owners—would always be there. In the Bay Area, there were more connections to Almanor, since it wasn't that far (a five-hour drive, compared to eight or nine from southern California, a drive which those couples could not believe we did). All of this had Dan saying to Donna, "We have to do this, how hard could this be?"

Dan even mentioned his interest in Almanor to Nora, who said they'd join him if he ever rented a place there. Dan booked a week at Almanor for summer 2010. It was an experiment, a foray, a reconnaissance mission if you will, to assess the lay of the land (er, lake), and to assess the possibilities. Dan rented two houses (and that's the trick; for multiple-family gatherings you need two houses), one on the water on the peninsula, in the country club, and another up the hill. The latter was a nice, large house, complete with game room, and capable of sleeping both families, as well as Katie's, who arrived midweek. The other house was...well, it

Lake Effect

was crap. Dan said that house was crazy, decrepit, maybe even dangerous. The kids thought it was haunted (by whom? Displaced Maidu? Drowned settlers?) and so nobody slept there. But that house was on the lake, and so formed a great base of operations for lake activities, which, as you've probably surmised by now, are the white-hot, beating heart of our lake experiences.

It had a dock.

The importance of this cannot be over-emphasized. Because had they not had a lake house with a dock, had they just had the nice house up the hill, had they just been hanging out on that house's deck in the evening (which I'm sure was very nice, was everything a cabin deck should be—spacious and comfortable, wooden, plenty of seating, barbecue, umbrellas, pines mere feet away), had they just been going to the community beach to swim, I daresay our lives today would not be the same. They would be slightly diminished, somewhat less than what they could have been. Sure, they'd be good—many lives are, after all, when all is said and done, when the pluses and minuses are finally tallied. But something would've been missing. Something you couldn't quite put your finger on, some vaguely unsettling thought that it could have been slightly better, that there was some missed opportunity left on a shelf, a road you didn't go down, a road that might have led to Shangri-La or Cibola, or Atlantis, or, if not all the way to one of those fabled places, then at least to a spot where you could see it shimmering in the distance, could affirm its existence. Because sometimes that is enough. Just to KNOW that such places are out there, that gorillas slowly pace the forests in the Congo, that orcas hunt in Arctic waters, that the Grand Canyon knifes into a layer cake of geologic history going back billions of years, that the current Pope believes dogs go to heaven. Maybe cats, too, but I'm less sure of that, myself, being a cat owner AND a dog owner. But because that house had a dock, many of us now know the pleasures of the dock; our lives are richer for it.

Tim Coonan

And no, Dan did not use the dock to tie a ski boat to, or fish, which are really the intended uses of a dock. But for one, the kids loved swimming from it; there was an inflatable swim structure in the little cove, one whose owners were apparently not there that week (many Almanor house are second homes or weekend getaway spots for their owners) and the kids took full advantage of this. Second, it dawned on Dan that there was really no reason to leave the dock. I mean, why would you? You had everything you needed right there, within arm's reach. Food and drink. Water to swim in. Good tunes on your portable speaker. And a beautiful lake sunset in the near future. So Dan DIDN'T leave the dock. He and his family and Katie's and Nora's stayed. Stayed until the shadows of the tall pines grew long, until the ospreys stopped hunting for fish, until the setting sun lit up Mt. Dwyer on the eastern shore of the lake. And then they stayed some more. Until the stars appeared and the kids started asking about dinner. We'll have dinner eventually, guys. And they did, but not until darkness had fallen on the lake and the houses and boats twinkled with lights, and you could hear the few boats out there with young people, because sounds seem to carry farther over water at night (akin to Alcatraz prisoners hearing the parties on the San Francisco docks at night). Thus began a precedent, observed to this day: you don't leave the dock until night has completely fallen. Every night of your week at the lake ends this way, the long and warm summer lake day leading up, without fail, to the cool of the dark night on the dock, to the blanket of stars over your head, to the friendly lights of the cabin up the slope. To the last good song playing on the speaker.

And it also set up the central daily tension of this lake vacation; just when DO you leave the dock? I mean, the kids have to eat SOMETIME (the kids, when young, accepted the late dinners because they had to; as they've grown older, they're not so accepting!). And you're certainly not planning on sleeping on the dock. One night Dan and Neal and Jim found themselves lingering on the dock, even though the families had headed up the hill to fix dinner, debating the merits of staying on the dock versus the price they'd pay for not joining their families up the hill. That was in 2010, and, to this day, that question has not been answered definitively. As you will see.

Dan also went looking for our family history that week. He (like me) was obsessed with finding the Hillyer house, our family high water mark at Almanor, to stir those memories and see if they held up in the light of day, under the summer sun 40 years later. He didn't find it. They rented a boat for a day, not a ski boat, but rather a mellow patio boat, similar to those Dad rented when we were kids (but greatly upgraded). Dan and the crew motored out of Big Cove and cruised down the peninsula, looking for the Hillyer house from the water, like Stanley searching for the source for the Nile. And like Stanley, Dan never found what he was looking for. Dan thought he'd recognize the Hillyer place from the water by its distinct, sawed-off A-frame shape. But he couldn't find it; turns out a pine tree had grown up in front of the house, one that was just a sapling when we stayed in the house 40 years before. Blocked the view of the house from the water, with the result being we wouldn't lay eyes on the Hillyer house for some years. Isn't that always the case, though? Things change, as the water slips past the boat and the years roll under you. Trees (and kids!) grow up, inevitably, obscuring the past from view. Finding the Hillyer house remained unfinished business for Dan.

The boat trip did start another tradition, though. Dan knew he wanted to upgrade from their crappy lakefront cabin of that year, and what better way to look for decent rentals than from the water? To this day we do that. Because there's always a better house out there, one with a bigger deck, a better view, a quieter cove...

And as it turns out, Dan and his family had visited Tahoe earlier that summer, enabling a direct comparison of the two lake experiences. Almanor won, hands down. Quieter, less expensive, less crowded, easier access. And Dan found that you only needed one good restaurant for a week at the lake, and they discovered one that week (more on that later). All the elements of what now comprise our Lake Almanor trips were there that week, maybe in rudimentary form, awaiting perfection (and eventual obsession): hanging out on the dock, and drinking, and listening to music, and going to that one good restaurant.

When you think about it, Dan's return to Almanor was not unlike General Douglas MacArthur's return to the Philippines during World War II. MacArthur had vowed to return to those islands, from which he had retreated early in the war, and return he did, walking ashore through the surf with his retinue, a camera-worthy moment if there ever was one. Except, as my dad points out, there was a dock nearby that he could have easily used, and probably should have, but it wouldn't have made for quite as dramatic a photo. And here's where Dan differs from MacArthur. Dan would have used the dock.

Chapter 12

The Russians Are Coming

If Dan's trip in 2010 was a reconnaissance mission, then our trip to Almanor in 2013 was a full invasion. Dan invited us all, and I mean ALL, to meet at Lake Almanor in July. "All" included Terry and his family, who flew from Florida to Sacramento, rented a big carry-all, hit Costco, and headed to Almanor. We affectionately call Terry's family the Russians, and they include Katja, born in Russia, and her five kids. One Thanksgiving they all drove out to California from Florida, which is a pretty big do, and it was greatly anticipated by the rest of the family. "When are the Egyptians coming?" Dan's son Kevin, then five years old, asked. "They're the Russians, honey, not the Egyptians," Donna, his mom, explained.

The Egyptians/Russians arrived at Almanor along with the rest of us, including my parents; it was their first visit to Almanor since 1978. Dan had rented TWO houses, next door to each other, on the southeastern shore of the lake, across from the peninsula, and my parents stayed in another up at Hamilton Branch, up at the top of the lake, well away from the mayhem that ensued at the other two houses. Smart move by Dan.

Now, the houses we rented were at the bottom of the lake's eastern shore; they were not on the peninsula, where all our previous stays had been. Which ignited (or maybe just started) a perennial discussion on the ideal place to stay at Almanor. That eastern shore had views of Lassen Peak, and great sunsets, because it was west-facing. But it was also subject to the afternoon winds which kick up every day, causing the dock to rock so much in the afternoons that the gangplank access bobs up and down making it difficult to deliver drinks to the dock, especially after you've had a few drinks yourself. The house had a large lawn, on which we set up a volleyball net for the kids, and in one corner of the deck was a hammock, discovered, and used thoroughly, by Neal.

2013: tucked under Dyer Mountain, this house was ALMOST everything we wanted in a lake house.

The week was slightly controlled chaos. Each family took a turn fixing dinner, which was a formidable task, since there close to 30 of us (I'm actually not sure how many of us there were—this would require some statistical sampling, such as the US census uses). But the fridge was packed. The bar was stocked. The deck was littered with beach towels and beach chairs and floaties, which the afternoon wind drove over to the next dock, requiring a rescue mission to retrieve them. Two phones ended up in the lake, one requiring a 40-mile trip by Terry to replace. The other—Carrie's—was quickly thrown in uncooked rice by Christina (owner of the first drowned phone), and that managed to dry it out. The phones were not the only casualty of the week. On a drive back from a peninsula restaurant, Dan and Donna realized that their son Kevin was not with them. He had gone to the restroom at the end of the meal and come out of the bathroom to find

Lake Effect

the whole group, all 30-some of us, gone. Luckily the wait staff knew he had been in our huge group and called Dan's friend Michele, who got ahold of Dan. Dan and Donna hurried back to the restaurant to find Kevin drinking cokes at the bar. He was fine, the bartender and wait staff taking good care of him. And Dan's concern at the at point was that our parents not find out about this egregious parenting fail, but it was too late for that. The word had gotten out quickly, in the initial flurry of text messages trying to ascertain which car Kevin might be in. Ah well. We are not our parents, and never will be.

    Mom and Dad found themselves on a Lake Almanor vacation, but without having to do any of the planning or logistics —and isn't that blessedly the case, as we grow older? Our kids take the lead; mine already do. I took my girls to Italy in 2016, and we each fell into our roles for the trip: Carrie was the leader, deciding where we'd go each day and even which room in the museum we'd hit up; Bridget was the affable trip member, with great day-to-day-attitude, and me? I bankrolled the trip and went along with everything. Thus my parents did not have to "direct traffic" on this Almanor trip; they did not have to rent the patio boat or arrange for horseback riding or fix sandwiches for anyone but themselves. Or put towels on the railing to dry or apply sunscreen to already-sunburnt skin. They joined us on the boat ride and Dan even enticed them out on the dock one evening, though they needed some help negotiating the rocking gangway. And, perhaps in a nod to old times, we all attended Mass with them at Our Lady of the Snows Catholic church, a fairly new place of worship right on the peninsula, and one that didn't require a drive into Chester (and, unfortunately, no post-Mass reward trip to a soda fountain). Mom and Dad even took their turn at hosting dinner one night, at their cabin up at Hamilton Branch—they certainly had the most experience of any of us at feeding a large crowd.

    That week the blueprint was drawn up for all future

Almanor trips. A patio boat rental one day, with a lakefront tour of peninsula homes and a search for bald eagles, coupled with donut rides for the kids (and adults) wherein the boat driver (particularly if it was Neal) tries to dump the inflatable donut rider (particularly if it was Nora). Hand signals (such as SLOW DOWN) apparently meant nothing. Various combinations and permutations of cousins riding the donut together, and attempting to stand up, and utterly failing to do so when Jim was driving. Another day renting jet-skis, satisfying the need for speed. Daily dock-centered activities: beach chairs, laying out, kayaking. Moving the patio umbrellas (and their heavy-as-hell bases) down to the dock to, in theory, protect our mayonnaise-colored Irish skin. Cousins delivering beers to the dock, and nobody getting carded.

On Golden Pond? Mom and Dad enjoyed a lake vacation during which they didn't have to do anything.

Jim gave us an astronomy lesson, taking advantage of the clear night skies and lack of light pollution. In urban and suburban areas light from streetlamps and houses greatly reduces the number of stars you can see, but not so in the mountains—you can even see the magnificent Milky Way on moonless nights. Talk about some perspective. As I teach my middle schoolers, our Sun lives on one of the outer trailing arms of this spiral galaxy, and it's 100 thousand light years just to get to the other side of the galaxy. And if you tried that you'd probably be sucked into the gaping maw, the event horizon, of the black hole that exists at the center of our galaxy (and almost every galaxy). Black holes totally confound the middle schoolers, just freak them out. I won't even go into the fact that the universe is expanding at an ever-increasing rate. Dark energy. Dark matter. Gravity warping time, like a bowling ball dropped into a net. Whew. Astronomy becomes philosophy at some point. On subsequent trips Jim would take great shots of Almanor skies at night, including an elusive comet, even getting up in the middle of the night to do so.

The eastern shore gifted us great sunsets. Jim gave us astronomy lessons and great night sky images.

And every evening ended on the dock, and the beach, with a series of great sunsets, each one different from the previous night's display. This in turn started a family photo trend with sunsets behind family group members, and groups of cousins. Christmas card material for sure.

We found we really didn't have to "do" that much. Nobody fished. Nobody water-skied or went wakeboarding There were a couple trips to Lassen, by Terry's family (bound and determined to see the local sights) and Jim and Killian (looking for that good hike). But Dan's approach to the dock as center of activities was similar to his MO for football tailgaters (and we are tailgate veterans, being lifelong Notre Dame fans): when you're throwing a tailgate party, you don't leave your own tailgater to visit anyone else's. It's just not done. You stay at home. Complete your assignment. The dock was the meeting place, much more relaxed than a living room or even a kitchen table. Somebody was always at the dock. In fact, usually there were MANY at the dock. It was like a family wedding weekend, but one in which you had the time and opportunity—thanks to the dock—to catch up with everybody. The geometry was circular. Beach chairs arranged in a circle, widening and contracting as folks arrived and left. No splinter groups. One big "circle of trust," as my Uncle John, the social worker/counselor, would say. The dock days started mid- to late morning (though some of us were up much earlier, as you'll see) and stretched lazily into the early and mid-afternoons. And they didn't stop there. Every one of the days was a long day's journey into blessed night. Thank goodness for the long, warm summer days in the Northern Hemisphere, itself a product of the Earth's axial tilt. Boy, if the Earth's axis had no tilt (the wobble in which gives us regular ice ages), and we had no seasons, and thus no Northern Hemisphere temperate forests, with their bears and wolves and lynx, well, what fun would that be?

And although we stayed on the eastern shore of the lake, a few of us wanted to find the old Hillyer house, on the peninsula, again. Because our memories of that place, that stay, that ONE WEEK, were so strong, for us it had become some sort of Platonic ideal of a lake vacation, our brief, ephemeral summer version of a Camelot. Such is nostalgia, that wonderful tonic that allows you to remember the past as maybe better than it really was, and to conveniently, even mercifully, forget the difficulties and stress you went through. It's very useful. Without it we'd probably never have more than one kid. Or go through a house renovation.

Lake Effect

Every day the dock reached—if not exceeded—the OSHA-established limit on occupants.

We decided to drive our parents over to the tip of the peninsula, for old times' sake, and Dan and I quietly hoped we'd find the Hillyer house again. It was there, and we were able to recognize it, because of the distinctive squared-off A-frame design—and because of the sign on the road that said "Hillyer." Of course, we wanted to walk around to the lake side of the house, but we didn't want to offend or disturb the current residents of the house, whoever they were. Our best bet was to send Katie and Nora up to knock on the door, accompanied by Dennis' three-year-old Seamus; that was the most inoffensive and likable combination we could put together, short of my parents. It worked. And here's the funny part—the current residents were the Hillyers themselves. They were my parents' age and had bought the place in the mid 1960s. Kids and grandkids had filed through it over the years. They got along with my parents famously that day, as people of that generation do. Turns out they had removed the firepole some years before, for safety reasons, and it was stashed in the garage. The place was light green now, instead of dark stained pine, but the house was basically the same, no huge additions or remodels (and believe me, some of the homes on the peninsula are now downright palatial). The Hillyer house had retained its old-school, somewhat vintage quirky cabin charm, and the fact that we met its gracious longtime owners was a homecoming of sorts.

Dan was not content to let this moment pass; he decided to recreate our famous deck-Christmas card shot of 1975, 40 years later. Because if you think I'm obsessed with the past, you'll find that Dan is every bit of the hide-bound traditionalist that I am. Dan asked if we could take a picture on the back deck, and Mr. Hillyer even took the picture for us. Dan made us line up in the same order as the original picture and wore a Little League all-star hat (his son Tommy's) for the recreated shot, just as he had in 1975. Perfect. Except for one detail. Terry wasn't there. Terry, who has often presented a moving target in life, was somewhere in the Lake Almanor area that day, but didn't make it over to the peninsula with us. Ah, such is life, isn't it? You can't really recreate the past; something is always missing. But we remember it fondly, all the same.

Chapter 13

Morning on the Deck

The Hillyer house hadn't changed substantially in 40 years. Nor had we, of course.

It was on that trip that I discovered—or rediscovered—a distinct Almanor pleasure, one that I enjoy perhaps more than anyone else. And part of the reason I discovered this is because I get up early. It's an affliction, I know, but now it's just part of me. I get up earlier than anyone else. During my school work week I get up at freaking 4:30 a.m. I'm drinking coffee and reading the papers (LA Times and New York Times) shortly thereafter, and by 5:20 (I know, that's freakishly exact, as if, like T.S. Eliot's J. Alfred Prufrock, I'm measuring my life out with coffee spoons) I'm doing my morning yoga. Which in itself is a daily reset for me, allows me to begin the day with movement and with quiet contemplation

Tim Coonan

and intention. In fact, I've concluded that the daily yoga allows me to love people more, which, in this day and age of rage and bad behavior, of January 6th and voter suppression, I find increasingly difficult to do. So yes, that's part of my morning routine. Maybe the best part. Shower, walk Duke, breakfast, and I'm in the classroom by a little after 7. Because I need all the prep time I can get!

So I found myself up early at Almanor. Earlier than anyone else. First light type of early. Because, for one, why waste first light when you're in a beautiful place? This is something I've learned through experience. And you don't understand this when you're young, or your body won't let you; kids are wired to, if anything, stay up late and sleep in late. Appreciation of sunrises and sunsets comes with age, when you realize that there are, in fact, a limited number of sunrises and sunsets you will experience, and you don't want to let one slip by. So I don't. I've taken students on weeklong trips to outdoor education camps, where you 're immersed in nature for a week, in an experience that can be life-changing for kids. Still, I could convince NONE of them to join me for sunrise. No matter. They'll stumble upon that pleasure when they're older.

There was no one else up that first morning at Almanor. And the house was packed with sleeping bodies, as those houses tend to be when we gather (the house "sleeps 12"? More like 20. Easy.). I quietly made a pot of coffee, but there was no way I could enjoy a cup anywhere in the house; the deck was the only option. And it turned out to be perfect. The house's deck was wrap-around, and three sides of the house had deck space adjacent to the tall trees, and views of the lake and mountains. I sat on a patio chair, coffee cup on the deck railing, and the forest started mere feet away from where I was sitting.

Almanor is not a busy lake to begin with. It is not crowded with water-skiers or jet skis or wakeboards, the way some other lakes are. And sure, they're there, but not till later in the day, and it still doesn't feel crowded, even then. Active, maybe, but not crowded. At dawn, especially on a weekday, there are a few anglers trolling the shallows or heading out to their secret spots on the other side of the peninsula. But that's it. And there's no wind at dawn. The wind will come up later in the day, especially on that west-facing shore of the lake, but at dawn it's still. Just me and the trees. And that's what I like.

Fear of missing out (FOMO) gets me up early.

The trees. Such marvels of evolution (and isn't every organism a marvel of evolution? Sculpted by natural selection to do well in its particular environment). Photosynthesis, that miracle of turning sunlight into carbs, photons into food, occurring in its thin needles, which are perfectly adapted to both capture sunlight in chlorophyll while remaining thin enough to survive winter's snow and ice and high winds. A northern hemisphere thing, for sure. Broadleaved, deciduous trees have to drop their leaves before winter, to avoid losing whole branches to winter's crushing snows. Not conifers, baby. We're open for business, gonna photosynthesize all year round. As I teach my students, the sun is the ultimate source of energy here on Earth, and we're only here because plants have learned to turn sunlight into energy. We certainly can't do that, and so we're in debt to everything that is green.

Tim Coonan

And this photosynthesis is taking place hundreds of feet in the air. Which is amazing to me, for several reasons. First, as I mentioned early on, this is because of competition—whoever can put their leaves (or needles) high enough will bag those photons, and go on to reproduce, to send those genes to a next generation. Competition within and among species has fashioned this natural world of ours, in an evolutionary arms race. Second, it amazes me that these conifers have evolved to be really good at standing hundreds of feet high. I mean, it's not as though they're made of steel, or rebar-reinforced concrete. Or even bones (and our bones would definitely not support us if we were hundreds of feet tall). All these trees have to support them are their plant cells. And the older cells, the heartwood, become so hard that they can support that tall, straight tree, helped by the massive root system which nails it to the earth.

If the trees could talk...they probably tell me to try sleeping in.

And the plumbing! Trees need to send water all the way up from the root system, where it is collected, to the needles, where water is required for the chemical reaction of photosynthesis. Conversely, the sugars produced in the needles by photosynthesis need to be transported back down, to all living portions of the tree. To me, it's an evolutionary miracle

Lake Effect

that xylem cells carry water hundreds of feet up a tree through capillary action alone, no pumps. Phloem cells bring those sugars back down. The phloem cells harden and become protective bark; the xylem cells harden and become heartwood.

On the deck in the morning, I'm amazed by all of this. It's a good time and place to contemplate such things. And I'm overwhelmed by the beauty of it. All those furrowed tree trunks, reaching high...I look up, at the sky, framed by the trees, and I see green. Millions of needles. Several conifer species dominate the forest at the edge of Almanor, and lakeside builders have done a good job of keeping that forest there, of wedging houses in among the citizens of the forest. The long-needled trees are ponderosa pines (the ones I loved in Flagstaff), one of several species of what are called yellow pines in the West. Their deeply furrowed bark has a bit of orange in it, and if, on a warm day, you stick your nose into the furrows, between the plates, you'll pick up hints of vanilla, or maybe butterscotch. This tree, and its long-needled cousins, the Jeffrey pine and lodgepole pine, are pretty much what you think of when you think of a pine tree.

The shorter-needled ones are white firs and have a classic Christmas-tree appearance. If you look up at the sky beneath their canopy, the twigs on which the needles sprout come off at right angles to each other, like thousands of crosses. The other common tree down near the shore of the lake has unmistakable shaggy reddish-brown bark—the incense cedar. Instead of needles, it has flat sprays of scale-like leaves.

This mixed-conifer forest has its other denizens, too, and on a quiet morning you will see, and hear, them. The noisiest are the squirrels and the jays, both of which will visit the railings of the deck, looking for the peanuts the kids left for them yesterday. Jays, in particular, really don't give a fuck. They are bold, and raucous, and smart—corvids (jays, crows and ravens) are undoubtedly the smartest of the avian set. The jays at the lake are the magnificent Steller's jays, indeed stellar with their rich blue body hues grading to jet-black on their heads and jaunty crest. Social, and successful around humans in these woods, they'll fight each other for the nuts we leave for them. I know if see Steller's jays, I'm in the mountains. In southern California we have their oak and chaparral-loving cousin, the scrub jay (its white throat appears to be "scrubbed") and Flagstaff has the colonial pinyon jay, inhabitant of the pinyon-juniper forest in those parts. Higher elevations yet will get you views of the gray jay, notable camp robber (they all are, and they're really good nest predators as well; scrub jays are known to follow bird biologists to nests and steal eggs).

Raucous Steller's jays and subtle mountain chickadees will join you for morning coffee on the deck.

There are other, more subtle, more quiet forest residents who flit through while you're sitting there, sipping your coffee and watching the light grow on the lake. You'll hear them first. They are small birds, gleaners as we say, going after the insects that live in the furrows of the bark and on the branches and needles. These guys come in what they call mixed foraging flocks: individuals of several species, darting through the forest, softly keeping in contact with each other, most likely so they can warn each other of predators—those accipiter hawks with sort stubby wings that specialize in catching songbirds on the wing: Cooper's hawks and sharp-

Lake Effect

shinned hawks. I have never seen those on my deck mornings, but the smaller birds know they are there. The mountain chickadees with their black cap and eyestripe and chick-a-dee-dee call, dark-eyed juncos with their outer tail strips and black, executioner-hood heads. Delicate white-breasted and red-breasted nuthatches; you can hear their soft "honk" before you see them. They are known for working a tree trunk going down; brown creepers work them going the other way: up. Sometimes you'll hear the rat-a-tat-tat of woodpecker, a white-headed woodpecker or a northern flicker, not satisfied to merely glean; they drill-hammer their own holes to get at the insects. And sometimes they drill-hammer the eaves of the houses, much to the consternation of cabin owners.

The calm of the early morning lake itself, hours from when the winds rise up, or the jet skis and ski boats head out, is appreciated by waterfowl. You can see mother mergansers, red head feathers trailing behind as if in fast flight, actually in a slow paddle among the docks, brood of ducklings in tow, fishing the shallows. The young'uns dawdle and then hurry to catch up, little webbed feet paddling furiously under the water. Further out on the lake are the western grebes, with their dramatic black-and-white plumage, gathering in large groups and skittering away or diving, for a surprisingly long time - where'd they go?—at the approach of a boat.

Then there are the ospreys. Formerly—and aptly—known as fish eagles, they wheel above the lake, looking for their prey. You can see the result of a successful power dive, as the osprey rises from the lake with a good-sized fish in its talons, off to a ponderosa perch to pick it apart. Another marvel of evolution: ospreys carry their fish lengthwise, instead of crosswise, to be more aerodynamic. But the bald eagles here are the royalty of the skies. There has been a breeding pair on the peninsula for a good number of years, and Dad even captured one in flight on his movie camera in the 1970s. Folks on the peninsula track the eagles' whereabouts; our friend Michele can tell us what side of the peninsula they're nesting on in a particular year. Nothing beats seeing a bald eagle in flight, and their continued presence here, and throughout the West, is testament to the effectiveness of the Endangered Species Act. Bald eagles were listed as endangered in 1978, but by 2007 had recovered sufficiently to be removed from that list.

Merganser families take their morning constitutionals near the docks; western grebes dive for their breakfast farther out.

For me, this morning scene is meditative, is mindfulness of the highest degree. I can focus on the conifer forest and the coming and going of its inhabitants, binoculars at the ready for a closer look at the regulars, and at the occasional rarer species, such as the bright-hued western tanager. I can —and have—rid my mind, for the most part and at least temporarily, of the troubles, the imperfections of this life. I am just in the here and now, seeing and hearing the lake morning. This place, the deck at Lake Almanor, does it like no other for me. And I find that I need to lose myself in the natural world every once in a while, if not on a regular basis. Hiking has this effect, and occasionally surfing (provided those two activities are not too crowded). It's the Japanese concept of "forest bathing," immersing yourself in the natural world. It's good for the soul. And it probably has a lot to do with why I became a wildlife biologist.

Of course, you can take this a step further. If coffee on the deck is good for you, then yoga on the deck is REALLY good for you. I have a "home yoga practice," as the yoga studios say: I'm too cheap to pay money to attend a crowded yoga class. But you gotta do yoga in a nice, peaceful setting. At home, my back deck suffices, but early morning yoga on the deck at Almanor is the ultimate. Surrounded by the forest, lake and mountain view in front of you. I have taken to doing yoga every other morning at Almanor, and it's the primary image and the experience that I take back home with me. The happy place, the thought of which sustains me until Almanor week rolls around the following summer. Dan's daughter Claire has ramped this up: she has done yoga on the dock, and even on a paddleboard on the lake.

The average age of the Dawn Patrol is weighted...up there.

I'm typically up at first light, and I'm eventually joined by some other early risers, though none as early as me. Dan's son Tommy had stomach problems for several years, problems that forced him to eat on a strict regimen, especially if he was planning on working out later. So sometimes I'd see him up early, too, quietly eating some breakfast well before the rest of the house awakened. Dan himself would also appear, as well as my dad, as the light grew brighter and the

Tim Coonan

deck warmed up. By this time, I'd be on my second cup. However, there would be no sign of anyone younger than 40 for many hours to come. Which is as it should be. Those young bodies and brains are wired to stay up late and sleep in late. Not mine, for sure. And I can't imagine I was ever that way! Give me that morning light.

My dad's presence at Almanor in 2020 led me to another unique experience. My mom had just passed, a few weeks earlier, and we convinced Dad to join us at the lake, scene of so much significance for our family. He was up for it, though the 9-hour drive was tough for him, and he was no longer getting around well. But he loved it. He loved sitting on the deck and watching the Steller's jays hop along the rail, going after nuts. By this time his daily walking had been reduced to walking the deck at home (10 laps, three times a day, each lap faithfully called out by this engineer) and he brought that habit to the lake, loving his deck walks there. I drove him down to the tip of the peninsula, with its 270-degree view of the lake, and of Lassen Peak, and he loved that, too. While we were down there we saw a bald eagle land in one of the tall pines. We passed the Hillyer house on the way down to the peninsula, and reminisced about that exceptional week, and the firepole, and Dennis, and that dock.

The thing is, that trip he and I were assigned to the same room (as the two single men), but there was just a queen bed in there, and no real floor space for an inflatable mattress. I decided to give my dad his privacy, so he could adhere to whatever schedule he wanted. So...I slept on the deck. I rolled my sleeping bag out on an outdoor couch. Every night I fell asleep under a blanket of stars, the bowl of the sky above me, framed by the dark silhouettes of the trees, and every morning I woke at first light, or before. Pure heaven. Surprisingly, there were no mosquitoes, and though our trash was visited by racoons (going after the avocado pits) I never saw or heard them. If I woke up during the night, it was to a pretty peaceful scene. I remember reading about a couple in Manhattan Beach who slept in a bed on their covered porch every night. If I owned a house at Almanor, I'd be tempted to do that. Until about October.

Chapter 14

Biological Diversity

The diversity of California's land, its landforms, shall we say, is well-known, and even legendary. Everyone knows that in California you can ski in the morning and surf in the afternoon, if you'd like to (though I'd reverse that; surfing is better in the morning, before the afternoon winds come up. And, uh, I don't ski). And everyone can name the dominant, and diverse, areas of California: desert and mountain, coastline. Most know of the Central Valley, as well. But it's actually even more varied, and diverse, than that. Where I live in Ventura, the southern California coast is characterized by a Mediterranean climate and ecosystem, where hot, dry summers and cool, wet winters produce chapparal vegetation and mild weather, which is featured yearly and prominently on Rose Bowl parade and game telecasts. The islands off that coast, the Channel Islands, where I spent the bulk of my career as a wildlife biologist, are also a world unto themselves, but little known outside California. In a similar vein, the Almanor area is much more diverse than people, including me, realize at first glance, and is a source of fascination.

This gets at the concept of biodiversity: that areas (or even the world itself) has a variety of living things in it, whether those things be species or plant communities or habitat types (we'll get to that). And, of course, more biodiversity—more things—is thought to be better. Back in school, I learned that places with more biological diversity (with more species, in particular), were thought to be more stable, could handle perturbations more, as if all those

Lake Effect

threads in its food web made it less vulnerable to disturbance. This may be so, and upcoming climate change will certainly test this. But to me, the concept of biodiversity is important for another reason. We don't want to lose any of it, because it's what we've been gifted, or entrusted with; it's what we've been tasked with preserving. Certainly the National Park Service, for whom I worked, takes this affirmative responsibility seriously. National parks are tasked with making sure there is no impairment of park natural and cultural resources; park values must be left "unimpaired for future generations" (that's directly from the NPS Organic Act, or establishing legislation, in 1916). A laudable goal, and one the agency has largely achieved. It's an important concept. As the great conservationist Aldo Leopold said, the first rule in intelligent tinkering is keeping all the parts. I was lucky enough to contribute to saving some of these "parts" at the Channel Islands, where we were able to bring an endangered species, the island fox, back from the brink of extinction. Put one in the W column. But I digress. The point is, California has significant biodiversity.

Some of this biodiversity results from California's long north-south axis, one effect of which is considerably different climate along that gradient. It's all about sunlight (in fact, life here on Earth is pretty much all about the sun, as I tell my students), and the direct sunlight at the equator bathes that area in so much more heat and light that do the indirect, angled rays of the sun in more northern climes. Add to that the more extreme effects of seasons at higher latitudes (so much more light in summer, so much less light in winter), and you get a wide range of climate zones in California. Linked to the sun are the precipitation patterns which directly, and obviously, determine the plants and animals which can take up residence in an area. The direct light at the equator warms that tropical air, which rises, with significant moisture; that vacuum, as it were, sucks in cooler area from higher latitudes, creating convection cells, huge conveyer

Tim Coonan

belts of moist air movement from tropical to temperate. These huge cells of movement are dragged by the earth's east to west rotation, resulting in phenomena such as the trade winds, which swept European ships to the Americas. In California the prevailing west or northwest winds bring moisture from the Pacific Ocean, which condenses as it cools down at high altitudes in the mountains, and then falls as rain or snow on the western slopes. There are much drier conditions on the eastern sides of the mountains; the air masses have dropped their moisture loads on the western slopes, creating a rain shadow (and Death Valley lies in a triple rain shadow, an extreme example of this).

This interaction of climate and topography in California makes for unusual and interesting biota. Take redwoods and sequoia trees, the tallest and largest trees on earth, respectively, and perhaps appropriately viewed as relicts of a prehistoric wetter climate. Coast redwoods rely on the moisture provided by cool maritime fogs that hug the northern coast, so much so that the big trees do not venture inland much at all. Their inland cousins, the giant sequoias, also require water in the form of soil moisture, and so only occur on the western slopes of the Sierra Nevada in areas where there is adequate runoff from summer rains at higher elevations, trapped by bedrock occurring near the surface. Giant sequoias were much more widespread throughout the West during the Pleistocene, when the climate was wetter and cooler. And while both these species are unusually long-lived (up to 2,000 years), they can't hold a candle to the bristlecone pines of California's White Mountains, where the slow, deliberate growth of this stunted and gnarled high-elevation pine allows individuals to live for 5,000 years or more. Those guys have seen it all, haven't they? Other California weirdos, products of its varied climate and topography: the funky Joshua tree of the Mojave Desert, and the Dr. Seuss-like giant coreopsis of the Channel Islands.

So, what about Almanor? The area has warm, fairly dry

Lake Effect

summers (continental air being drier than maritime air) and cold, wet winters, with most precipitation occurring as snow at its elevation, which is about 6,000 feet. And geography gives the Almanor area a surprising amount of biodiversity itself, because, well, it is not all one thing. You may think of it as "California mountain stuff," but it's more complicated than that, and deserves a finer cut. This begins with the mountain range in which the Almanor region lies. Though you think it might be in the Sierra Nevada range, that fabled massif of John Muir and Yosemite, Sequoia and Mt. Whitney, it's not. The Almanor area is actually in the southern-most extension of the Cascades, that volcano-heavy range which drifts into northern California from its base in the Pacific Northwest, in Oregon and Washington. In fact, the Almanor area lies at the confluence of several, shall we say, influences.

There are so many ways to characterize an area, in terms of natural features. There is the concept of geomorphic provinces, where the underlying geology (including the morphology, or landforms) holds sway; the biology arranges itself atop this patchwork quilt of topography. Other scientists favor the concept of "habitat type," but to me, that seems unnecessarily animal-centric. I recall one visiting professor in grad school facetiously saying that all plants were either "bird roosts" or "bird food." Or you could simply use "natural regions," a term which covers a lot of ground (literally) and is maybe the most intuitive. In any case, Almanor sits at a biological intersection. The Sierra Nevada massif stops its northward march in the area; the volcanic Cascades invade from the north, and even the more coastal Klamath ranges reach over this way from the west, as does the Modoc Plateau/Great Basin Desert from the northeast. This intersection, or confluence, of biologic provinces brings species from each into the area, increasing its diversity. Edges are like that. It's the intersection of a pretty big biological Venn diagram.

Tim Coonan

Lassen National Volcanic Park sits in this intersection, and is a great, and well-recorded (since national parks are well-studied) example of this high biodiversity. At a little over 100,000 acres, it's not that big, as far as national parks go (Grand Canyon National Park weighs in at 1.2 million acres), yet Lassen harbors over 1,000 plants and vertebrates. Thanks to its peaks, which include 10,000-foot-high Lassen Peak, the park has the wonderful altitudinal zonation in plant communities that you get as you go higher. Higher elevations are colder, of course, and my fifth graders learn this is because, even though it's closer to the sun, the air molecules are sparse at the lower pressures of high elevations, and they just don't bump into each other enough to heat things up. Most of the park is below 6,500 feet, and is mixed conifer forest, characterized by the usual suspects we know from the woods surrounding Lake Almanor: ponderosa pine, white fir, and sugar pine. Climb above this, and you're in the red fir forest, where that species is joined by white pine, mountain hemlock and lodgepole pine in thick, dense forests dotted with openings. But even these higher-elevation species have their limits, and as you go higher, the growing season gets shorter. When you get to 8,000 feet, the trees peter out, the cold and the dry—moisture being tied up in snow—being too much for them. This is the subalpine zone, and above this is the alpine: at 10,000 ft, near Lassen's peak itself, trees are no more.

Just as the trees sort themselves out along this altitudinal gradient, so do the animals. Down low (and low here is relative; it's still 4,000 to 6,000 ft in elevation) you'll find ducks and geese, even loons, on the lakes, river otters, and the occasional beaver dam on streams, those ecosystem engineers creating their own reservoirs. Mule deer and black bears roam the edges and meadows. The weird mountain beaver, more closely related to squirrels than to the true, dam-building North American beaver, inhabits moist forest areas but doesn't range too high, since hibernating is not in

Lake Effect

its bag of tricks (nor is dam building).

Up higher, the red fir forest has fewer species, but is home to those which prefer its dark, thick woods, long winters and snow-covered forest floors. The rare and secretive spotted owl roosts here, sharing the woods with the equally secretive marten, a sleek carnivore related to weasels. Its cousin, the fisher, might be here; they're so rare and secretive that their status is unconfirmed. Other mustelids (skunk-like carnivores) include American minks and long-tailed weasels, and the rare Sierra Nevada red fox also haunts these environs. You might see a northern flying squirrel, the original BASE jumper, gliding through the canopy. A select group of songbirds inhabit the red fir forest: mountain bluebirds, golden-crowned kinglets, gray jays; black-backed woodpeckers and Williamson's sapsuckers drill and tap for insects on the boles of the trees. Rare goshawks and blue grouse live here, too.

Go higher still, into the subalpine, above 8,000 feet. The trees here, the whitebark pine and mountain hemlock, look different, because they are stunted by the cold and lack of moisture and heavy loads of snow. Still, life finds a way, as Ian Malcom said in *Jurassic Park*. It's amazing to me that animals make a living up here, but they do. Snow-white snowshoe hares ping-pong through the many openings in the brush. On barer slopes, where wildflowers and low vegetation take hold in crevices between the rocks, several rodents do well: one large, the woodchuck-like marmot, and one small, the tiny pika, "farmer of the tundra." And the subalpine has its complement of breeding birds, including red crossbills, pine siskins and the high-elevation specialist rosy finch.

Down the eastern slopes of Lassen's mountains, toward its eastern border, the influence of the arid Great Basin Desert is evident in the stands of sagebrush and bitterbrush, rabbitbrush and needlegrass as you proceed east and downslope and away from the moister western slopes of the park. This semiarid habitat is the domain of jackrabbits and

Tim Coonan

coyotes, deer mice and kangaroo rats. A world away from the also desert-like alpine habitat on Lassen Peak.

And despite all this biodiversity, all this meeting of biological provinces and altitudinal zonation, all these plants and animals sorting themselves out by habitat type, there are still some things missing. Things that were there once. Big things. Things that would make all the difference to the wilderness character, to the naturalness of the area. Things that may, or may not ever, come back.

Chapter 15

Big, Fierce Things

We have never done well with big, fierce animals. And for good reason. What self-respecting Neanderthal would have tolerated the presence of giant cave bears in his or her domicile? Or perhaps more accurately, what self-respecting cave bear would have tolerated a Neanderthal family? This is not only competition, but is what the biologists call interference competition—one competitor kills the other. And then there is hunting; early humans may have had a hand, or a spear, or even an atlatl, in the demise of many of the wonderful and varied Pleistocene megafauna: mammoths, giant ground sloths, dire wolves, saber-toothed cats, American lions, short-faced bears. I'm amazed that these creatures once frequented the Los Angeles area, where some were trapped and preserved in what we now call the La Brea Tar Pits. Those fossils have been dated from 40,000 to 9,000 years before the present, and so many of those species would have been hunted or defended against by early humans, who were in the Americas by maybe 15,000 years before the present (and who may have peopled the Americas by following the "kelp highway" down the Pacific Coast). As satisfying as this thought is—it allows us to finger humans as a cause of catastrophic environmental change with obvious implications for the current climate crisis—it likely did not happen. More evidence now points to natural climate change, a decrease in global temperatures, as the cause of extinction for these Pleistocene species, as opposed to human hunting.

Be that as it may, humans have not fared well with large

Tim Coonan

predatory animals, and have largely attempted to extirpate them from areas, with remarkable success. Take the grizzly bear, perhaps the most fearsome large animal on our continent today (now that saber-toothed cats and American lions are gone), and one that is on the California state flag. It no longer exists in California (indeed, California is the only state with an extinct animal on its flag). Grizzlies once occurred throughout the state, in almost every habitat save the southeastern deserts, perhaps 10,000 strong, feeding on deer and elk. The great bear inhabited the Almanor area as well; early settlers of Susanville recorded sightings, somewhat fearfully, in the 1860s. However, the California gold rush and explosion of ranching created conflict between humans and grizzlies, and the bears' tendency to kill cattle (as well as the occasional human) put a target on their backs. Grizzlies were gone from the state by the 1920s. I doubt they'll ever be seen in the Almanor area, or in California, again. Except on state flags flying above municipal buildings.

There's more hope for another species, however. Out of all the big fierce creatures, wolves seem to occupy a special place for us. They are mythical, ghostlike, appearing here and there. Maybe we can relate to their methodical, team-like approach to hunting prey, the fact that they live in packs, extended family groups (like we do), with an alpha couple at the head, wolf pups cavorting outside the den. They're related to our favorite domestic companion, the dog (dogs were bred from wolves long ago) and exhibit doglike behavior which we recognize. And they don't really pose any direct threat to us. They're pretty timid, when it comes to humans. There are only two documented cases of wolves killing humans in North America in recent times – and one of those is suspect (might've been a grizzly bear). Wolves are nonetheless the stuff of legend, from the "big bad wolf" to *Game of Thrones'* dire wolves (and yes, I was pissed off when Robb Stark's wolf died at the Red Wedding).

Wolves once ranged widely in northern climes, from

Lake Effect

North America to Eurasia, even inhabiting Japan and the British Isles, as well as deep into Mexico. Early hunter-gatherer peoples largely revered and coexisted with wolves, but the advent of agriculture and herding changed all that. Wolves were, and are, extremely efficient at killing domestic livestock, and agricultural societies had no tolerance for such. These practical reasons were bolstered by the advance of Christianity, which taught that humans were the masters of the natural world, and not a part of it. By the Middle Ages northern European and Russian literature was replete with depictions of wolves as rapacious, wanton killers of both livestock and humans. Wolves were eradicated through local efforts in Europe, with the result that wolves were gone from the British Isles, Scandinavia, and Central Europe by 1800.

European settlers brought their wolf-hating views to North America with them. Wolf persecution began immediately: almost as soon as they stepped ashore; both the Plymouth and Jamestown colonists established wolf bounties, and the war on wolves only grew as colonists expanded westward across the continent. The boom of the livestock industry in the 1870s and the disappearance, due to market hunting, of the West's huge bison herds spelled doom for wolves, as they turned more and more to preying on domestic livestock. State and local authorities set a goal of extermination of all predators, including wolves and coyotes, and in fact, this became a national goal, as well. In 1915 the federal government established the Division of Predator and Rodent Control, and paid hunters went after the last of the wolves, even in areas where there was no conflict with livestock (the U.S. government still kills predators; in 2020 the ironically named Division of Wildlife Services killed, by its own reckoning, over 400,000 native animals, including 381 wolves and 62,000 coyotes—all to ostensibly protect livestock). By 1930, wolves had disappeared from almost all the lower 48 states, thanks to a systematic program that represented, in the words of one biologist, "the longest, most relentless, and most ruthless persecution one species has waged against another."

The demonization of wolves predominated throughout the early part of the 20th century, particularly in rural areas, where up to 90% of farmers and ranchers, to this day, still disapprove of wolves (in contrast, the majority of western state residents favor wolves and wolf restoration). Favorable attitudes toward wolves—by some—began to appear in the 1940's, boosted by Aldo Leopold's proposal to restore wolves to Yellowstone, and detailed field studies such as those of Alaskan wolves by Adolf Murie. Farley Mowat's 1963 fictional *Never Cry Wolf*, eventually made into a Disney film, and Barry Lopez' 1978 *Of Wolves and Men* propelled sympathetic views of wolves into popular culture. Enlightenment occurred, even in the halls of Washington, D.C. Passage of the federal Endangered Species Act in 1973 turned attention to the plight of imperiled species, and the gray wolf was classified as endangered under that law shortly thereafter. They had finally received full Federal protection. But wolves were all but gone from the lower 48 states.

As my dad would say, nature abhors a vacuum. There were still hundreds of wolves in Michigan and Minnesota, despite years of bounties and wolf control, and across the Canadian border there was a shitload. Maybe 50,000 wolves or so. Resilient as all get-out, they had fared well through control programs that had even involved poison. And it was as if they saw ads in the trade magazines (Wolf Illustrated, Wolf Vogue) advertising free territories south of the international border. No passport or visa required, just make your way here and get a deed from the local land office. It wasn't exactly the Oklahoma land rush, but wolves did in fact pour over the border. Border Patrol had no shot at stopping them.

And so they came into the U.S., as hopeful as any immigrant, just with larger canine teeth and more fur. Wolves crossed into Montana in the 1970s, into the Rockies, and in 1986 a pair denned and produced a litter there, the first in

Lake Effect

over fifty years. The original immigrants must have written home with glowing accounts of the area—maybe they even wired money back—because wolves kept coming. An impetus grew to reintroduce them, to reestablish them, in the heart of their former Western range, in the regional ecosystem with the most gravitas, the most significance: Yellowstone.

I will tell you, Yellowstone is absolutely revered among us parkies. It's the core of the Greater Yellowstone ecosystem, the most intact temperate zone ecosystem on Earth. It's the mother park, the first national park established (though, like the exact origin of baseball, that is debatable; Yosemite can also make that claim). Yellowstone's origins involve an 1871 scientific exploration expedition camping near the area's remarkable thermal features, its geysers, and a campfire discussion concluding that the area need to be preserved for all time. A great story, and one that is largely true. The expedition's glowing reports of the area convinced Congress to protect the area as a national park a mere six months later (have you ever known Congress to act so quickly?), an act that not only protected the geysers but also the northern Rockies ecosystem on its high-elevation plateau, "and the wildlife therein" (I love that quote from the NPS establishing legislation) including its elk, deer, bighorn sheep, bison, grizzlies, and wolves. Well, okay, the bison, grizzlies and wolves were gradually, or not so gradually, all but eliminated over time, but, um, at least their habitat was still there. And it was still there in the 1990s, when the federal government proposed reintroducing wolves to Yellowstone and central Idaho, a proposal so controversial it required an environmental impact statement and three years of public meetings before approval was granted.

The U.S. Fish and Wildlife Service really wanted reintroduced wolves to stay put, which is no mean feat, so family groups (packs) of wolves were captured in Canada and released in Yellowstone using "soft release" methods: the groups lived in acclimation pens for several weeks before being set free to do their thing. And do their thing they did. Established and defended territories. Had pups. Hunted elk and deer. And in so doing, they changed the Yellowstone ecosystem, for the better. Yellowstone's numerous elk, unchecked and at an all-time high prior to wolf reintroduction, had camped out in the area's steams and riparian areas, absolutely hammering them. Wolves changed all that. Elk numbers dropped and elk behavior even changed, perhaps due to fear of wolves; elk left the riparian areas, which recovered. Willow and aspen grew. Beavers built dams. Bison increased. Coyotes decreased. And wolves...continued their peregrinations, their restless wanderings: Montana, Idaho, Wyoming, Washington, even Oregon. By the late 2000s they were at the doorstep of California.

In 2011, the unthinkable happened. A wolf stepped foot —er, paw—into California, the first to do so in almost ninety years. And all hell broke loose. The wolf was now-famous OR-7, who returned to Oregon and established a pack in the Rogue Valley, but not before exciting conservationists and putting the fear of God into ranchers. And not before, apparently, telling other wolves about the bounty of the Golden State. A pair of black wolves (as if this couldn't get any more mysterious) appeared near Mount Shasta in 2015, prompting a flurry of wildlife biologist activity. Trail cameras returned pictures of seven black wolves, the pair and their litter of five. Wolves were back breeding in California. The pack mysteriously disappeared after that. Did they pull up stakes and leave? Were they the victims of foul play? Unknown. But they weren't the only pack in California.

In 2015 a trail camera recorded a picture of a gray wolf in Lassen County, which in itself is delicious irony: the last known California wolf was killed in Lassen County in 1924. Biologists, as they are wont to do, investigated further; genetic testing of blood samples revealed the wolf to be a female born in the Rockies in 2014, who had traveled over 800 miles to the Lassen area. And by 2017 she had hooked up with a male from the Rogue Pack, forming the most successful pack to date in California. The Lassen Pack, as it is known, is one of four now in California, and has had litters annually, increasing to fifteen wolves (and counting). The pack has ranged over a 500-square-mile area in Lassen and Plumas counties, a territory which, amazingly (to me) skirts the eastern edge of Lake Almanor! In fact, wolves from the Lassen Pack have been spotted on the outskirts of Chester. Were they headed in to town for supplies? Ice? Firewood? Ice cream?

Pups from the Lassen Pack frolic (as pups are wont to do) in 2017 (U.S. Forest Service photo).

To me, this is perhaps the most remarkable of all the natural history aspects of Lake Almanor (okay, the volcano is pretty damned impressive, too). That, as I sit on my Almanor deck drinking morning coffee and gazing east, there are wolves, not that far away, on the other side of Dyer Mountain, doing wolf things. Coming in from a night of hunting deer (and the occasional cow). Mating. Raising pups. Nuzzling noses and sniffing butts and otherwise engaging in social behavior. They are here, in the area, and are a massive ecological presence, and yes, they are pretty much back from the dead. Ghostlike. Back from two centuries of persecution, because, as it turns out, we couldn't stamp them out. Life finds a way. The Lassen Pack certainly offers us a modicum of hope in these cataclysmic times, no?

Chapter 16

Dock Talk

The wolves were out there, somewhere beyond Dyer Mountain, as I walked toward the dock, but they were the furthest thing from my mind. It was summer 2019, and we had rented a house on the eastern shore of the peninsula, down near the tip of the peninsula, in fact, near to the Hillyer place and the lake house of our friends, Joe and Michele O'Neill. Daily (nightly?) Dock Talk was beginning. But this wasn't a formal event; there was no official starting point, no call-to-order, no quorum needed. It just organically happened every day, as bright day slid into cool evening and night. The dock was heavily used all day, of course, as folks sat in beach chairs, read books, sought shade under the two umbrellas brought down from the dock, or lay on towels in the sun. Folks came and went, stepping carefully over sunning bodies and stand-up paddleboards and paddles, asking, is this chair taken? Very light conversation during the day. Folks swam or dipped after a while in the sun, or carefully mounted the massive float raft toys. Some paddled on the standup paddle boards or kayaks. I suppose it was an Almanor day like any other Almanor day.

Not that the week had been without incident. Carrie and I almost didn't make it there, for one. And when I say "didn't make it", I'm talking about the mortality sense of that phrase —we had been in an accident on the drive up, and my car had been totaled. Carrie was driving us through Stockton, on 1-5, when a car came fast up behind us, aggressively weaving its way among cars. I was thinking, this is not good. It cut

Tim Coonan

in front of us and tried to pass another car, to the right, but didn't make it. Clipped it. Spun around and slid to a stop in front of us, perpendicular to the road. Carrie hit the brakes but couldn't stop in time, and we hit him, on the left front of our car. I didn't think it was that bad—the airbags didn't even deploy—and Carrie drove the car (still drivable!) over to the shoulder. By this time the young man driving the car was heading our way, accusation in his eyes. He never got to us. He was intercepted by what I now call our guardian angels: three middle-aged Black women not about to put up with his bullshit. "We saw you!" They yelled at him. "You were driving like a maniac!" By this time the CHP had arrived and began to arrest the driver, who was obviously on drugs. Carrie was pretty shaken up, and one of the women—God bless her!—came over and said, "It's not your fault, sweetie," and give Carrie a big hug. The driver's companion, a young woman, used my phone to call her dad, and I'm glad I didn't hear that conversation. While the CHP was managing the accident scene a drunk driver rolled through and hit a CHP vehicle, and so the CHP arrested him, as well. Neal was an hour ahead of us and doubled back to pick us up (God bless him, too). The tow truck driver told us that Stockton was the meth capital of California (how did I not know that?). We had the car towed to an auto repair place which was not open, it being Saturday, and I parked it outside the fence and tossed the keys over, paddleboard still on top (Neal already had a board on top of his car). I thought, naively, that the place might have it fixed by the time we drove back down the following Saturday. Nope. By Wednesday they had finally looked at it and determined it to be totaled. My brother Dennis, who had not planned on an Almanor trip that summer, drove to Stockton from San Jose and gathered all my things from the car, including the paddleboard, and brought them to Almanor later in the week (God bless Dennis, too). Carrie still cringes when we drive through Stockton, but some good came out of it, as she points out: that day we

Lake Effect

managed to get both a meth addict AND a drunk driver off the road.

My car wasn't the only casualty of the week. Katie's daughter Colleen fell down the steps at the lake house—some of them were uneven—while unloading things on the day they arrived, necessitating a trip into Sacramento and an emergency visit to a dentist. And, I missed a great wildlife moment, which occurred IN the rental house. I had gone to bed (early as usual). Nora had been wrapped up in a throw blanket on a couch, when her daughter Grace noticed something crawling on the blanket. It was a baby bat. I'm not sure how much of a freak-out Nora had—she's not exactly a wilderness gal—but Grace wrapped the blanket around the young bat and released it outside. Now, you might figure, where there's one baby bat there might be others, as well as a mom, around, but I inspected the house the next day and could find none, nor any spots where bats could access the house. But bats would be featured again on this trip, as you'll see.

The dock was blessed relief after all this craziness. On this particular evening, by the time the shadows had lengthened and the light had left the dock and was only full on the eastern side of the lake, on Dyer Mountain, Dock Talk had begun. Now, the basic configuration of Dock Talk is circular. Everyone is in a circle, a rather amorphous one, which just grows bigger as more folks arrive. And I must tell you, at this point, that Dock Talk may have some roots in what my Uncle John called "Deck Talk": folks sitting around on his and Pat's deck in Worcester, Massachusetts, talking about "how the pigs eat the slop," as John so delicately put it. John, the professional counselor and social worker, also called it a "circle of trust," somewhat facetiously (John's tongue is permanently in his cheek), and so this circular Deck Talk may have partially influenced Dan's choice of the name "Dock Talk" and its basic circular arrangement. May have, as I said, because we are loathe to give Uncle John too much credit for anything; after all, as we tell him, he married into this family and is not a blood relative, and never will be. But if he wants to think he has had some hereditary influence on Dock Talk (which by now has far surpassed Deck Talk in scope and influence), I suppose there's little harm in that.

The circular arrangement is also essential for the music. See, Dan (or Jim or Neal) has a portable speaker wirelessly connected to Dan's (or Jim's or Neal's) phone. And it's not as though they choose the music; we all do. And here's where the circular arrangement comes in handy. Everyone down there, in turn, picks a song to play and adds it to the queue. Then you pass the phone to the person next to you. Your song will be played, eventually (hopefully not while you're on a bathroom break), although sometimes the queue is a full lap or more behind the music. And sure, you could choose any song to play, but let's just say the crowd appreciates songs that are significant, that speak to the moment, to the folks who are there. Yes, you're playing to the crowd, in effect. In fact, you can even play songs targeted at one person in particular. For example, if you play Eric Clapton's "Wonderful Tonight" it will remind Nora of her wedding day when she danced with Neal to that song. Neal proposed to Nora to the sounds of Springsteen's "Little Girl I Wanna Marry You," and that will tug at Nora's heart, too. I have played James Taylor's "Sweet Baby James" and Karla Bonoff's "The Water Is Wide" to remind Bridget and Carrie of when I used to sing those to them as lullabies. Shameless, I know. There is some pressure here, to choose a song that is significant, and yes, acceptable (Neal is probably the harshest critic). Acceptable songs include most from the classic rock genre, though the younger generation has introduced, successfully, I might add, some country western, many of the younger crowd being Stagecoach festival goers. And the acceptable music narrows down considerably, as the evening progresses.

Over the years, there has become a familiarity to this

scene. On the dock by now are a number of coolers, but Neal's large soft-sided one is the mainstay, from which Coronas (with lime wedges) are dispensed, and Neal is oh-so-ready to pour you a margarita, as well. Calls go up to the house, requesting more ice or more beers, as supplies are depleted. Dinner is a long way away; it has been deferred (like so many Covid graduations). Still, Dock Talk is hungry business, and the food of choice is tortilla chips and artichoke dip, from Costco.

Music, drinks, food…sure, you say. But what about the actual Talk of Dock Talk? What do you actually discuss? The weighty issues of the day? Prospects for world peace? A solution to global warming? Is there an agenda, with items to discuss? In order? Maybe this is actually a family meeting, to decide what to do about an errant family member. Worse yet, maybe it's actually an intervention for that family member (and that thought has actually occurred to me. If Dan ever says to me, "Sit down, Tim, we'd like to talk to you," then I'm getting the fuck out of there as fast as I can).

As far as what we talk about, I can't help you there. I have no idea what we talk about, and I've been at Dock Talk for a long time. Maybe the best way to explain it is like this: on Seinfeld, George Costanza (one of the greatest, and most pathetic, sitcom characters of all time), concocted an idea for a television show that was, as he said, "about nothing" (which was wonderfully self-referential to the Seinfeld show itself). Dock Talk is kind of the same thing. It's a talk about nothing in particular. In fact, it's not really a talk. Which is not to say that it's insignificant. We have celebrated graduations on the dock, especially for those whose graduations were waylaid by Covid, and one of the next generation has said her wedding will be on the dock (in which case we will need a much larger dock). So, unlike a meeting, which has outcomes, and action items, Dock Talk is more of journey than a destination. It's a gathering just for the gathering's sake; there's no art to it. It's unscripted. It's Mass without the liturgy, the missalette. It's a birthday party without the birthday kid, a family wedding without a bride and groom. In fact, it's really just a setting, an environment, albeit a very pleasant one, in which family and friends can sit around and do...nothing in particular.

Now, Dock Talk has been a slightly different experience at the different houses we've rented. Across the lake, on the eastern shore, the house we rented had wonderful sunsets and a view of Lassen Peak, but also stiff afternoon winds that set the dock a rockin'. For a couple of summers we had a house on a quiet cove at the top of the peninsula, where lake waves and winds rarely disturbed us. We named it "Coonan Cove" (and, again, you won't find that anywhere on Google Maps) and enjoyed watching the dogs at the house across the cove, and noted the woman doing lake yoga every morning at the dock at the cove's tip. Our Dock Talk there may have been the most excitement that quiet cove had ever seen. This particular year, our peninsula house faced straight out to Dyer Mountain, no cove; the lakeside line of docks in almost perfect order up and down the peninsula. And pretty damned peaceful down at our end.

This particular evening happened to be one of a full moon, and it rose over the southeastern edge of the lake, to the right of Dyer Mountain and to the left of a little point with a tall ponderosa pine, and lit the place up like Christmas. Nobody left the dock for any reason; why miss a full moon Dock Talk? About a half hour after sunset, as dusk deepened but before night descended, the bats appeared. They had been doing this all week, at this appointed time, and they swooped and dove right above the lake, hunting insects. The air became thick with bats, and some flew right over the dock, at close quarters, maybe a little too close for some. But not too close for Carrie, budding biologist and adventurous experiencer of nature. She decided to paddle out among the bats. She headed out on a standup paddleboard, and I was sure the bats, most likely little brown bats, would occasionally hit her in their feeding frenzy. But bats are darned good at what they do. Evolution has honed their echolocation skills so they not only can avoid flying into objects in inky blackness, but also find and catch minute, moving insects. Carrie disappeared into the darkness as she paddled out, and I could see the silhouettes of bats in front of and behind her. She later said they brushed her occasionally, or maybe that was just the air churned up by their soft, furred wings as they deftly dealt with this object moving over their feeding ground.

The deck at Tantardino's rivals my parents' dining room as a place for family dinners—only with more tequila shots.

Darkness finally settled on the lake, broken only by the full moon and its reflection in the water, now calmer as the breeze died and the boats motored slowly into their home docks. The Dock Talk crowd thinned out. Somebody left to make dinner. The music narrowed in scope to Springsteen and the Eagles, with the final stretch comprising Eagles Dock Talk "top five," which include "Peaceful Easy Feeling" and "Last Resort." There may have been some shots of tequila. Then, by cell phone light, beach chairs and towels were gathered up, bottles and Solo cups went into coolers, and the last participants, the rearguard, as it were, carefully negotiated the gangplank from dock to shore (for some unknown reason a little shakier now) and made their way up the lawn to the friendly, welcoming lights of the house. But the music didn't stop when the Dock was vacated. As Neal and the hardcore remnants walked up, you could hear Tupac's "California Love" issuing from the speaker Neal was carrying. It's the song played at Dodger Stadium when their closer, Kenley Jansen, came out of the bullpen to ice the game. And it signals the end of Dock Talk, and another W in the books.

Chapter 17

Dinner Will Be Late

One night that year, late at Dock Talk, a guest appeared out of the darkness, and made her way down the gangplank to the dock, with a bottle of Patron in her hand. She was hailed and warmly welcomed, and not just because of the Patron. It was Gina Tantardino, one of two sisters (Trish being the other) who owned Tantardino's restaurant on the peninsula. Tantardino's had become one of the required elements in our Almanor experience, and sometimes we ate there as many as three times in our weeklong stay. As Dan says, you only need one good restaurant during the week, and in Tantardino's we had it.

Gina and Trish were raised in LA, as we were, and were even products of Catholic school there, as were we. Their dad moved up to Lake Almanor in the early 1990s and opened the restaurant, which quickly became a local, and not-so-local, hit. Their pizzas are flat-out good. In fact, they may be some of the best in California. I know, I know, "there's no good pizza in California." Especially if you hail from New York or Boston or Chicago, or the apparent epicenter of the Pizza Belt, New Haven. I will grant you that. No real argument. Still, pizza is pizza, and it's delicious, and there's good pizza in a lot of places.

In LA, when I was in high school, we'd go up to Brentwood and get pizza at La Barbera's on Wilshire, absolutely the best in LA. It was run by two Italian-American brothers, and when they closed the restaurant, we were rudderless, as far as LA pizza went. And their absence only made the legendary pizza better. Years later, and I mean decades, I hired a woman named Susan as a biological technician for the island fox recovery program at Channel Islands National Park (and believe me, that's a whole 'nother story). Turns out that Susan was dating, off and on, Joe LaBarbera, son of one of the former owners. Believe me, when I heard that, I was beside myself. Where are the brothers now? Are they ever going to open another restaurant? Can I get the recipe for dough, sauce? Turns out they had moved up to San Luis Obispo and had thought about opening a place, but they never did. The brothers did all the cooking on family camping trips, but that was it. I didn't get the recipe for pizza dough or sauce (and I really didn't expect to!), but Joe was kind enough to give me a recipe for bowtie pasta with broccoli, which ultimately became a family favorite. And I got a La Barbera's t-shirt, which is now in my relic t-shirt drawer, full of other such holy items, which only make it out for ceremonial purposes.

Tantardino's pizza might actually rival LaBarbera's, and I'm not the only person who raves about it. Makes me wonder if there was any overlap in LA between the two. Had Gina's dad, or Gina and Trish, eaten at LaBarbera's? Whatever that history, Tantardino's got it right. And why wouldn't they, with a name like that? And if you ever met the Tantardino sisters, you'd see why. Tough, Italian-American, they ran that restaurant with authority and purpose. Dan cultivated a friendship with them (as he did with the local realtors, smart man) and in turn Gina and Trish looked forward to our yearly visits. Our group—as many as thirty—would take over their back deck, one long table, kids at one end and adults at the other (in the middle were the tweeners, like Colleen and Bridget, who could go either way). From the table I could see Trish working the bar inside, visible through a window on which sat Dodger shot glasses – the sisters had brought their love of the Dodgers north into Giants country. As she did every year, Gina served us with intent and efficiency, happy

Lake Effect

to see us, and always giving Kevin crap for being left behind at the restaurant years ago. She relayed her orders to the kitchen through a window, on which sat a beer she occasionally sipped from. And at some point during the meal, a round of Patron shots would arrive at the table. In Dodger shot glasses. They were sometimes ordered by us, and occasionally by one of us who wasn't at Almanor that year (which is a fine tradition, if ever there was one: the Coonan who wasn't at Almanor owed everyone else shots). And sometimes Gina and Trish would order the round, because they joined us for the shots, Trish stepping away from the bar to do so. Now that is service. Any joint where the owners drink with you is a fine establishment. Toasts were made to Tantardino's and Lake Almanor and to the yearly visit, and as the years passed, more of the kids joined us for the shots, having finally reached drinking age.

And the food. Sometimes all I really want is to sit outside at Tantardino's and eat, as the sky darkens above the pines on the other side of the road, the strung patio lights get a little brighter, and the stars come out. First the salads: the Caesar is great, but the loaded wedge is outstanding. Yes, drizzle that iceberg wedge with blue cheese and bacon, and still call it a salad. The pizzas arrived and were huge; they sat on raised metal stands above the table, and may as well have been on plinths, the way Greek statues are set, according them the prominence they deserve. The heavily meat-laden, aptly named Carnivore. The Mediterranean, with its Feta and sun-dried tomatoes and red onion. Some of the kids favored Hawaiian, with its pineapple and Canadian bacon. Bridget liked the Spicy Santa Fe Chicken. As is always the case, more pizza was ordered than could be consumed in one sitting, and everyone eyed the takeout boxes which would be headed home. That pizza would not last long back at the house, certainly not past tomorrow's lunch, and maybe not even past tomorrow's breakfast. But there's always room for dessert, isn't there? And that's a proven biological fact.

Tim Coonan

Dessert was a chocolate brownie smothered in ice cream and chocolate sauce; sometimes the kids' end of the table would order several, unbeknownst to us adults. And then...well, the best comparison I can come up with is that of a shark feeding frenzy, after chum is tossed in the water. Or piranhas, after a tapir falls into the sluggish, muddy waters of the Amazon. Doesn't last that long and is surprisingly violent and intense. Those desserts rarely made it down to our end of the table (and the truth is, we'd order our own). By this time, we'd outlasted everyone else at the restaurant, and we filed out, pizza boxes in hand. Done till next year? Not hardly. We'd already booked a reservation for later in the week.

Gina's presence on the dock signified, in a way, our acceptance into Almanor culture, because Gina pretty much IS Almanor. It's a tight world, to be sure. Gina was married to the owner of the real estate agency we booked rentals through, and EVERYONE there knows Gina and Trish through the restaurant. Because, for one, it's a peninsula, and everyone passes by Tantardino's on their way into the Club. And with Gina's management and people skills, she could run the peninsula, the Club, all by herself.

The Club. Clubs are always so insular, so exclusive, aren't they? You have the folks in the club, and those who are not in the club. Then you have the renters, like us, who are club members, as it were, for one week a year. The folks in steerage on the *Titanic*. As a renter you don't have nearly the same status as the true club members, the folks who own property there. And you can tell. We've had true club members as neighbors for our week at the lake, and they wouldn't even look our way, much less say hello. They must hate the renters, with a new group of us arriving every week (the rental houses are booked all summer; it's actually pretty hard to nail down a good rental. That's why Dan befriends the realty office folks). Other owners have been gracious and generous: "Hey, we're heading back to the city for the week—use our dock!" Thank you, we will!

Lake Effect

Once you're in the Club (and I'm speaking geographically here; the Club comprising the lower half of the peninsula), you can use the amenities, such as they are. There is Rec Area 1, the biggest, halfway down the eastern side; it boasts a dock and boat ramp, swimming beach, picnic area, volleyball ball courts (which are more dirt than sand), and tennis courts, where the ubiquitous and baffling pickleball has taken over. It also has a bandstand, which, on summer Sundays, hosts the most popular social events of the summer (not counting the alcohol-soaked, illegal fireworks-laden Fourth of July, which we have always missed; I would not want to be on the lake that day). No, the Sunday events at the bandstand are rather tame in comparison, just a music-in-the-park type thing.

But I tell you, they are well-attended. It's usually our first event of the week, having arrived the day before, but we don't attend in the style the locals do, the style befitting this summer cabin, retirement-home place: in a souped-up golf cart. It seems every resident has one, if not two, and they are the predominant form of transportation in the Club, ferrying people to the local market, to the tip of the peninsula (dogs riding shotgun, eager for a short walk and a charge into the lake), to the golf course, of course, and to the weekly band concert. They are pimped-out and personalized, flying flags and donned in colors from favorite teams (and everyone is from somewhere else). Many have their golf bags permanently on the back. Groaning with the weight of grandkids, coolers and fold-up tables, the golf carts park in rows ringing the clamshell bandstand, and it's kind of like a drive-in movie, with folks watching the performance from the comfort, such as it is, of their cart.

The musical fare is primarily what you would call tribute bands, each danceable enough to attract a "mosh pit" of over-40-year-olds, some even line dancing. There is always a raffle, and folks come around selling tickets. Other residents throng the nearby picnic tables, with kids play pickleball and adults cornhole, smoke from barbecues rising up through the pines. Teenagers congregate, glomming together like magnets (they really have no choice in the matter; it's just the attraction of north-south magnetic poles) and pushing up against the boundaries of decent behavior. It's no wonder they stay up late and sleep late as well. Boundary-pushing is best done under cover of darkness, and they—we—are hard-wired to do so. Until they get to be as old as I am; then all you want to do is go to bed.

Look, you can arrive a little late to Almanor, but you can't miss the family photo at the concert in the park.

And us? We're in back, under the tall pines, behind most of the golf carts, sitting in lawn chairs and on coolers and blankets. Neal is pouring margaritas, and we're enjoying each other—haven't seen each other for a while—and enjoying the band from a distance. The girl cousins take a traditional walk together, catching up on everything that has happened since they were together last; which is everything they've already posted on Instagram. The cousins circle up and bump a volleyball around, volleyball pretty much being the family sport.

Nora, who is very perceptive about these things, is always on the lookout for the Lady of the Lake, a somewhat mythical and regal figure—think Galadriel, or Maggie Smith's character on Downton Abbey—who MAY be the oldest living resident on the peninsula. Heck, for all I know, she may be one of the three daughters for whom the lake was originally named. I believe she lives in one of the largest lakeside houses near Rec Area 1, and we have seen her at these concerts driving her golf cart, well, regally, if one can indeed do that. Our friend Michelle knows her (Michelle, like Gina, knows EVERYONE on the peninsula) and has introduced Nora. Nora is fascinated with her, and thinks there may be a heckuva back story, one involving prominent men at the lake, and has threatened to write an entire book on the subject. Which will no doubt make this book seem like a Beatrix Potter story.

The day dims, we pack up our coolers, and head back to our cars, well before the golf carts file out. It's not exactly the parking lot at Dodger Stadium after a game, but still worth getting a head start. The dock is waiting. And dinner will be late.

The fires at Almanor in recent years have made for spectacular sunsets. Which come at a dear price.

Chapter 18

The Smoke from a Distant Fire

I don't ever recall seeing wildfire, or its effects, on our trips to Almanor in the 1970s. There is no evidence of fire in the pictures or 8mm movie footage; no blackened trunks, no charred trees, no bare charcoal-ly slopes, no pall of smoke in the air, or plumes of ash. We never had to adjust our activities or take a different road to avoid a fire. The recent trips, however, have not been so fire-free, and this is true even of the "reconnaissance trip" Dan organized in 2010. On a lake boat outing that year Neal recalls seeing a slope southwest of Almanor that had obviously, and recently, burned. The presence of fire became ever more common as we returned to the lake each year. We all saw smoke from distant and not-so-distant fires on our way to and from the lake, as fires became more frequent, even commonplace, during the California summers.

It got a little closer to home, so to speak (and Dan would be glad to hear me call Almanor home) during our 2019 trip. That week the Hog Fire started burning to the northeast of Almanor, along Highway 36, which leads to Susanville and ultimately to Reno. The smoke drifted into the Almanor Basin, and a thin layer settled over the lake at night, before the daily winds dispersed it. Then one day the fire blew up. Stoked by winds, it exploded, creating its own weather. Our fireman friend Rob, visiting us with his family, recognized what was going on, as a huge anvil-shaped cloud of smoke and ash appeared to the northeast. The superheated air from the fire had risen, and when it did, it had cooled off at high altitudes and the water vapor in it had condensed into clouds. This was a "pyrocumulonimbus," a thundercloud created by the fire itself. The supercharged air over a volcanic eruption can also create these. Thunder, lightning, even rain can come from such clouds, and also vicious downdrafts which further stoke the fire, like a giant bellows. They are no good for fighting fire, that's for sure. For several days the Hog Fire sent these thunderboomers soaring to great heights, as the fire grew to a size which closed Highway 36. Dan had to go south of the lake to pick up Donna in Reno and the fire burned over a fiber optic cable, knocking out internet service to Chester and to the peninsula at Almanor. Other than that, we weren't much affected, though it made for dramatic afternoon skies and colorful sunsets. The fire grew to about 10,000 acres, a size which used to be considered large, and was eventually contained and controlled a month after it started.

Nora appears relatively unconcerned by the pyrocumulonimbus building up behind her.

Tim Coonan

Being a California resident, I was no stranger to fire. In December 2017, Santa Ana winds knocked over power lines near Santa Paula, east of my home in Ventura, sparking a blaze that raced westward, pushed by the easterly winds. The fire burned hungrily through coastal sage scrub and chaparral which hadn't burned in 50 or more years and was just ripe for it. In no time at all the fire was at the edge of Ventura, burning north of town through the hillsides. I wasn't worried. I had seen fire burn through the hills before, and in fact, there was a history of fire doing that in the Ventura hills. But this was different. Those strong Santa Ana winds pushed that fire into a housing development in the Ventura hills, a development dating from the 1970s which had never seen this type of thing before. Houses burned. The fire knocked out electrical power needed to pump water up the hill, hampering efforts to fight the fire. And the fire sent burning embers (it began "spotting," as they say) into my neighborhood, below the hillsides, igniting fires here and there. From my upstairs bedroom I watched houses burn, mere blocks away, and I thought about what I would take with me when the evacuation orders came. It was a pretty sleepless night. The fire burned everything in its path on its march to the sea. North of Ventura, the town of Ojai was surrounded by fire on three sides and was only saved by a fuel-break road which fortuitously had been cut above town, specifically for this eventuality. The fire burned into Santa Barbara County, pretty much all the way to the ocean, and crews fought to save the coastal towns of Carpinteria and Montecito. Two months after it began, it was out, having burned 1,000 structures (400+ in Ventura) and over 280,000 acres, making it California's largest fire ever, at that time. But the damage wasn't over. Runoff from torrential rains in January poured down slopes denuded of vegetation by the fire, overwhelming San Ysidro Creek in Montecito, overwhelming inadequate catchment dams, and sending a wall of rock and mud through Montecito. Over a hundred homes were damaged and twenty-one people died in those floods, which knocked out Highway 101 for such a long time that people in Ventura were commuting to work in Santa Barbara by boat.

The Thomas Fire left a pall of smoke over Ventura for weeks. Residents walked around town wearing N95 masks, well before Covid required us all to do so. This was a teachable moment for my middle-schoolers, and they learned about fire ecology in the chaparral of southern California. In that Mediterranean ecosystem, stand-replacing fires occur every 50–70 years, but historic fires were never on the scale of the Thomas Fire. Prior to modern humans occupying the area, there were no ignition sources during the fall and winter Santa Ana season; the only fires began by lightning during summer, when there were no Santa Ana winds to push them to conflagration size. Yes, we have changed the calculus on that. We now have fires starting by unnatural sources during Santa Ana season. Recipe for disaster.

But that's in southern California, not northern California. There are no Santa Ana winds at Almanor. And that doesn't seem to matter. There are high fuel loads (from years of fire suppression), tinder-dry conditions, and plenty of ways for fires to start, most of which involve electrical generating and transmission equipment. Wildfire completely dominated our stay at Almanor in 2021. Dan had booked us a house on the western shore of the peninsula, with views to the west, including Lassen Peak, and I enjoyed morning yoga with a view of that volcano. It was a long way down to the lake on that steeply sloped side of the peninsula, down so many stairs that Dad did not make it down to the lake that week. He still did his deck walks, on the deck around the house. It was his second trip to Almanor after Mom's passing, and he appreciated the significance of the place to our family. We ate pizza at Tantardino's. Went to the tip of the peninsula, noting the Hillyer's place along the way. Went for a couple of boat rides. Held Dock Talk every night (a little windier on that side of the peninsula). We even celebrated Carrie's graduation from St. Andrew's on the dock, since Covid had precluded a live graduation that summer.

Then, on Wednesday of that week, we started seeing smoke drift into the Almanor basin, from the west this time. A fire had started the day before, down along the North Fork of the Feather River (not MY North Fork of the Feather River! Please!). When I first heard about it, the fire was several thousand acres in size—dinky, really—and I thought, they got this. They're on it. It actually started just several miles from where 2018's devastating Camp Fire had started. I'm sure you remember that—the Camp Fire burned over the town of Paradise and killed 85 people, because of poor warning systems, and because there was just one road, one escape route, leading out of Paradise to safety. In November, not even fire season in the Sierra, if you ask me. Oh, by the way, a Pacific Gas and Electric transmission line caused that fire, which was then driven by easterly winds into Paradise, where it caused what they termed an "urban firestorm." I figured with that fresh in everyone's minds, they would throw everything they had at this fire which had started mere miles from the Camp Fire's origin, and it would be quickly contained.

It wasn't. Now dubbed the Dixie Fire, it was burning in the heavily wooded, steeply sloped canyon of the North Fork, and they couldn't really get at it. They even sent a train up there—it had long been a railroad route—with a special car built to spray water, but that couldn't get at it either. The smoke grew heavier every day, and we started checking the air quality values for the Almanor peninsula. By Thursday and Friday those had soared into the 400s; anything above 300 is considered very unhealthy. I stopped running, and we actually ceased all outdoor activity due to the pall of smoke. On Friday Dan arranged a lunch over at a diner in historic Prattville, on the west shore of Almanor, just to get us out of the house. Prattville was cloaked in smoke and ash stuck on

Lake Effect

the ground, like hail which hadn't yet melted. Dad ordered a huge banana split for lunch, and nothing else. Why not? The end of the world was nigh.

We left the lake on Saturday, as we always did, packing the cars tight with coolers and towels and recycling. Goodbyes at the lake, and then again at the In-N-Out in Chico, the traditional last hurrah for the cousins. But so much uncertainty hung in the air, like the smoke that dogged us pretty much all the way into Chico. Would Lake Almanor, the peninsula, have to evacuate? Could fire actually jump into the peninsula? There were only two roads that led out of the club. How much would this fire burn? Would we return to a charred landscape next year? I drove Dad and Carrie away from Lake Almanor, up Deer Creek, also heavily, beautifully wooded, with trees overhanging the rocky creek and historic split-rail fences lining the pastures and meadows. All very burnable. But CERTAINLY they'd stop this fire before it got this far.

They didn't. Or couldn't. The fire was 10,000 acres in size by Saturday morning. By Monday, 40,000 acres. Then it doubled over the next two days, and then doubled again. By the 23rd it was over 170,00 acres in size and heading for Almanor like a runaway freight train with the brakes out. The fire divided in two (almost like an army), one flank running to the west of Almanor, the other to the east, and it was all but unstoppable. It roared to the edges of Chester and Westwood, which were saved by very aggressive backburning operations, but the historic mining town of Greenville was not so lucky. Winds drove the fire through the town on August 5th, destroying three quarters of the town's buildings, including some which were over 100 years old. The west shore, the east shore, and the whole Almanor peninsula were evacuated. By this time the Dixie Fire was over half a million acres in size. It crossed the crest of the Sierra Nevada (the first fire ever to do so). It burned north into Lassen Volcanic National Park, torching much of it, including, somewhat ironically, an historic fire lookout tower. Ended up burning into five counties, and was finally contained in early September, by which time it had consumed close to a million acres, becoming the largest single fire in California history, and the most expensive to fight. All because, as the Forest Service determined, a "perfectly healthy" tree had fallen onto some power lines along the North Fork of the Feather River.

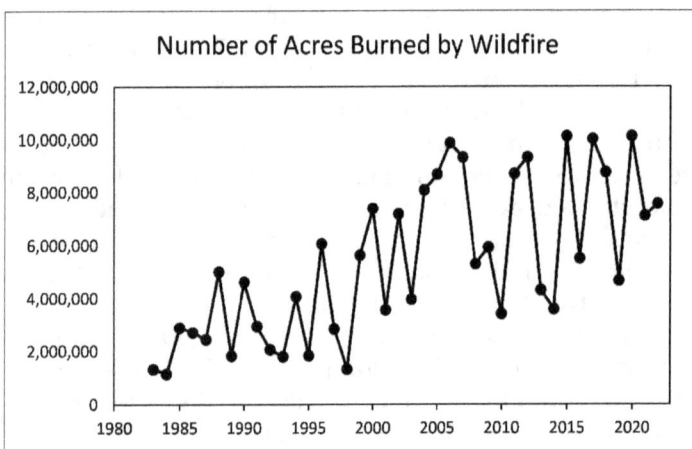

The number of acres affected by wildlife has increased over time (data from U.S. Forest Service Interagency Fire Center).

You know, I don't do well with change (as anyone close to me will tell you); I pretty much want things to be... immutable. I want the Notre Dame campus to be pretty much exactly as it was when I was there in the late 70s. The same with my hometown (with the exception of breweries; towns can always use more breweries). Some of this is my inordinate sense of place, crystallized in my 30-year National Park Service career. I want those national parks to be unchanging, to be, as they say, "left unimpaired for future generations." I want there to be glaciers and polar bears and

Lake Effect

African lions and Florida panthers, and island foxes, into the foreseeable future. I know all this is unreasonable of me, and the scientist side of me tells me that there is nothing constant but change, and that has been true since the Earth congealed some 4.5 billion years ago, and even since the Big Bang well before that. Still...the thought of not being able to return to the Almanor I knew, and have known, saddened me as I tracked the Dixie Fire's onslaught from afar. Because you want to be able to go home again, to return to those places which nurtured you and healed you over time. The thought of the North Fork scoured out by fire, or Deer Creek's streamsides choked with blackened, fallen trees, of bare slopes on Dyer Mountain across from the peninsula, of hiking to Lassen Peak through a gray moonscape...it all made me wonder what I would find on next year's Almanor's trip. And I knew that Almanor was not the only area experiencing such wholesale change.

Chapter 19

Out of the Frying Pan

In 1988 Yellowstone National Park burned, and it was a pretty big deal. Dozens of lightning-caused fires began burning in June and July, and the National Park Service let them burn, because the fires were natural, to a large degree (they had been started by lightning strikes), and the ecosystem was used to such fires—even required them—and so the NPS had what they then called a "let burn policy." In fact, an NPS fire ecologist famously said, of the fires, "Burn, baby, burn!" Something he later may have regretted, somewhat, given what those fires eventually did. See, the fire folks expected rains to come in July, as they usually did, but in fact that year they did not. So those fires took off, and by July they were being fought vigorously (by mid-July the NPS decided to suppress all fires). The fires eventually burned through close to 800,000 acres of the park, even through the Old Faithful area (threatening the historic Old Faithful Inn) and were only halted by early-season snow in September.

I visited Yellowstone the following summer, partly to see the effect of the fires and the aftermath, including any recovery. And indeed, nature was revving up again, postfire; there were lots of little green things, herbaceous plants, popping up among the blackened and charred stumps. These days the story of Yellowstone is not the 1988 fires, and any changes they wrought, but those fires were the first, I'd say, to draw national attention to large fire in the West. Fire has played an increasingly larger role in the West since that time. The federal government tracks such things, of course, and

Lake Effect

they're pretty good at it. Data from the National Interagency Fire Center, since they started tracking such things in 1983, shows that the number of acres burned every year has essentially increased five-fold in that time.

There is variation in this, of course, as there is in most natural systems. Much of this variation can be attributed to whether the western US is experiencing El Niño (warm, wet year) or La Niña (cooler, dry year) conditions, that difference dependent upon whether a big blob of warm water is in the eastern or western Pacific Ocean. When the warm water is near the western edge of South America, there's less upwelling of cold water, and the low pressure causes more rain and greater snow depth in California and the West; the storm track dips southerly during El Niño. The opposite is true during La Niña, when the storm track is more northerly, hitting the upper Northwest and tracking well north of California. El Niño itself refers to the Christ child, as these conditions are noticed near Christmas time in South America. While you might think the El Niño precipitation is good for quenching or preventing fires, the precipitation increases plant growth, and that vegetation is what the fire folks call fuel: it more readily supports fire that may develop when that vegetation (fuel) dries out, as it does every summer, or in a subsequent La Niña year.

These five- to seven-year El Niño/Southern Oscillation patterns are readily apparent in the graph of acres burned over time. But the number of acres is increasing over time, and that's not El Niño; that's climate change. As average temperature increases, vegetation/fuels tend to dry out earlier in the summer season and are hotter when the fires come. I took a Ranger course once at the Grand Canyon, and we learned what a wildfire thrives on: oxygen, heat, and fuel. To control or stop a fire, you need to eliminate one of these three factors; you need to "break the fire triangle." Unfortunately, climate change has made it harder to do this. Fuels are hotter now, and dry out earlier and for a longer period, which has extended fire season into the spring and fall, and even into the winter. There is no non-fire season, really. This stretches to the limit what our fireman friend Rob calls "resources": the crews and equipment needed to fight fire. Resources that could be sent to southern California to fight Santa Ana-driven fall fires have, in many cases already been committed elsewhere.

Interestingly, fires are also growing larger and faster under climate change. The drier, hotter fuels encountered by fires encourage their rapid growth, and as a result we see fires such as the Thomas Fire and the Dixie Fire just take off, growing rapidly, exponentially even, to sizes of several hundred thousand to almost a million acres. Thus, more fires are now what you might call megafires, or fire complexes (the latter term describing when two or more fires converge).

These climate-change factors are compounded by other ones. There are more ignition sources now, because there are more people in the West, living and camping; there are more things at risk, as more homes have been built in what they call the "wildland-urban interface": where cabin meets forest. My ex-wife and I owned a cabin (a log home, really) in the Pine Mountain Club area near Mt. Pinos north of LA. Even though brush removal and tree limbing were required of homeowners every year, the house sat among towering Jeffrey pines and white firs, and I had accepted the fact that one day it would burn to the ground. Living in the Western forests is now pretty risky—for residents and for the forest itself.

Climate change has produced more extreme weather, giving us high winds, storms, flooding, droughts...conditions that our modern infrastructure wasn't built to withstand, because we haven't seen this type of thing in historic times. Take electrical power equipment. Its failure to withstand high winds (and treefall) has been a weak link, and indeed has been fingered as the ignition source in several large California fires, to the point where utility companies have

Lake Effect

been held liable for the losses incurred during those fires. In fact, power companies are now pre-emptively shutting down power grids during peak fire season in fire-prone areas.

And it doesn't help that fuel loads in western forests are high, due to decades of well-meaning but short-sighted fire suppression efforts. The US Forest Service and fire ecology researchers are now looking at the effectiveness of thinning operations, which hold promise, but that seems to me a small-scale solution, one that could not be implemented to the extent needed to prevent conflagrations like the Dixie Fire. And any logging operation comes with its own environmental costs.

Then there are bark beetles. Which are native, apparently, and they do their thing: infect conifer trees. As they have done for millions of years. They target diseased or stressed trees, boring holes in the bark and depositing their eggs. A healthy tree can produce sap and push the beetles out before they lay eggs. But drought changes the equation here. A tree stressed by drought can't produce the sap required to evict the new tenants, and bark beetle infestations can really take off. The combination of drought and bark beetles is deadly, and recent drought in California encouraged beetle infestations which have killed over 100 million conifer trees in the state. Those dead trees are just perfect fuel for wildfire, as you can imagine.

Climate change has now put areas of California, even the state's unique biota, at risk of fire, a threat to which they had never previously been at risk. Take sequoias, those towering behemoths of the Sierran western slope, Pleistocene relicts, once widespread during cooler climes, they're actually dependent on fire, need it to crack open their cones and let the seeds out. They are, as they say, serotinous. Low intensity fire crawls through the underbrush under the big trees, causing the fallen cones to burst open and spread their seeds in the now nutrient-rich ashy soil. The mature trees are adapted to this and are rarely harmed; their bark is legendarily thick and fire-resistant, and their foliage is hundreds of feet high in the air, well above the relatively small flames. Low intensity fire is thus fine, and even needed. But high severity fire? That's another thing altogether. High-intensity fire can kill the giants, especially during drought, when the trees are drier and don't have as much fire-retarding moisture. And boy, has there been unprecedented drought. The 2012–2016 California drought, bad as it was, also came with higher temperatures than normally seen, creating a situation deadly for sequoias facing fire. The Castle Fire in 2020 killed as many as 10,000 mature sequoias, perhaps 10% of the Sierran population.

The following year, the KNP Fire Complex raced into the Giant Forest, in the heart of Sequoia National Park, killing as many as 3,000 sequoias. The National Park Service took extraordinary measures to protect the giants, including wrapping the boles of trees with fire-retardant wrapping. The following year, the NPS installed a sprinkler system around the Grizzly Giant, a revered sequoia in Yosemite's Wawona Grove. The tree itself is steeped in myth. Apparently, John Muir himself, demi-god of the Sierras, took Teddy Roosevelt camping under its tall canopy over a century ago, and the then-president was moved enough by the experience that he subsequently extended the federal protection of the young national park. The Grizzly Giant itself is over 200 feet high and over 2,000 years old. It's rare that individual organisms gain iconic status, but this particular giant sequoia has it (other named trees include the General Grant and General Sherman trees in Sequoia National Park). And the NPS had to put a damned garden sprinkler system around the Grizzly Giant to protect it from the nearby Washburn Fire (a decade of prescribed fire in the grove also helped save it from that particular fire; good on you, NPS—prescient!). Such measures, which seem almost desperate, may become routine, annual activities under climate change. Pass the Reynolds Wrap over here, will you?

Lake Effect

Pretty much every place I love in the west, Flagstaff, northern New Mexico, the Rockies, Yellowstone, Lake Almano—everything save the sparser lands of the desert and canyons – is now touched, tainted, even, by the specter of fire, if not by fire's recent onslaught itself. Fire runs amok, beyond the bounds of what nature in its pre-climate change mindset ever anticipated. We now schedule our Almanor trips earlier in the summer, before what used to be seen as fire season. But there is no non-fire season anymore. And I wondered what I would find at Almanor, after the Dixie Fire's march through the area, a march every bit as thorough and ruthless as Sherman's march to the sea.

Chapter 20

Is it Hot in Here, or Is it Just Me?

Now, my students don't know about Sherman's march to the sea. They would, of course, if my dad were teaching them middle school history, which he taught for a while, in addition to middle school science. My dad is a military history buff (along with my twin, Terry) and was known for teaching history off the cuff—he knew it so well, he didn't need a textbook, or even notes. No, my middle-schoolers don't know that stuff, but I do teach them one thing my dad felt strongly about: that geography determines history. Dad gave me a big topographic map of the U.S. for my classroom, a map he used in his, and I am always pointing things out to the kids: the geographic obstacles to westward expansion (the mountain ranges and deserts), the advantages of coastal locations, gradations of temperature and moisture. My kids can also tell you that oceanic currents and the trade winds enabled Europeans to find the New World pretty easily, thereby sealing the fate of North America.

My middle-schoolers also know about fire and climate change; I make damned sure of that. For one thing, they lived through the 2016 Thomas Fire, and need to know it was not just a one-off, but instead a harbinger, a sign of things to come. And they need to know what kind of world we're handing off to them. Here you go, guys, good luck with that! Let us know how it turns out. I mean, why sugarcoat it? It's perhaps the most important thing I can teach them, the one lesson I want them to know and realize and take with them before they go off to the uncertainty, the maze of high school instruction, where one has more choices (certainly more than we did, back in the day) and you can largely avoid serious science if you want to. Well, you're not avoiding it here, baby. Yup, you're doing a science fair project, because I can make you, and for the same reason, you might also have to be in a school play. These are mandatory middle school experiences.

And as much as I want them to know about the facts of global warming, I also want to them to know how to discern fact from fiction. That is, after all, what science is all about: uncovering evidence to see how the natural world works. Evidence, evidence, evidence, I hammer into them. And sources. Do you trust the source of your evidence? When the kids try to convince me of certain things—like the existence of mermaids, or that *Megalodon*, the extinct giant shark, is not actually extinct—I ask them, what is the evidence? What is the source of your evidence?

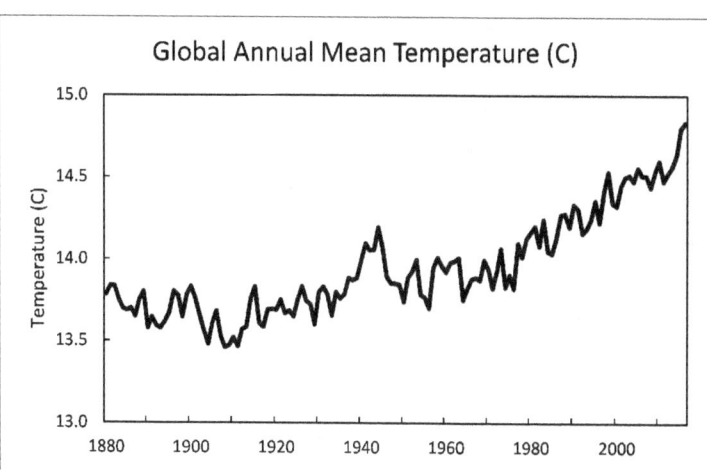

Average annual global temperature has increased steadily since the dawn of the Industrial Age (data from NASA Jet Propulsion Laboratory).

Tim Coonan

I want them to see directly the evidence for global warming, and believe me, it's out there. From reputable sources, too. I give the kids a basic spreadsheet, showing average annual world temperatures since about 1850. Graph it, I say. The kids graph it, and they do it well. In fact, they're pretty good at anything digital and visual. They play with the colors and the lines, the fonts. Make it pretty, I say. Now, what does it tell you? Man, if can get them to read a simple graph, a time series of ANYTHING over time, and tell me what it means, that's all I really want for them. My job here is done.

And this may be the most significant graph of scientific data ever produced. I'm sure you've seen it. It has just one temperature value for each year, one value that averages thousands of daytime and nighttime temperatures from all over the world, condenses all that variation into one yearly value for our Blue Planet. How can all that variation produce anything meaningful, you ask? Certainly, no trend could be discernible. Oh, but it is. Unmistakably obvious, this trend. You can put a trend line on it, if you want, even do a regression analysis to see if it's truly significant, or if your eyes are deceiving you, if the annual variation, the year-to-year swings in temperature, is the real story. But you don't have to. The best data displays are obvious, and this one is: global temperatures have increased since the late 1800s. They've increased almost 1.0 °C, which is about 1.8 °F (another distinction they need to know, the existence of two temperature scales), and that doesn't sound like much, but consider what has come with that increase already: a decrease in ice sheets and glaciers, a decrease in Arctic sea ice, an increase in ocean temperatures, and a rise in sea levels. This is pretty heavy-duty stuff. And where do these data (yes, in science, "data" is plural) come from? From NASA and the Jet Propulsion Laboratory! Kids, can you trust these data and their source? Hell yes, you can!

This leads naturally to many questions, the first of which

Lake Effect

is why the Earth has warmed. I show the kids another graph —in fact, I give them another set of data to graph themselves: annual CO2 values from measurements taken at the top of Hawaii's Mauna Loa, since about 1960. Hey, look at that, it matches the temperature rise for those years! In fact, scientists have extended that CO record back in the 1880s (and beyond, actually) and the values match the temperature increase nicely. Now, what conclusion do you draw?

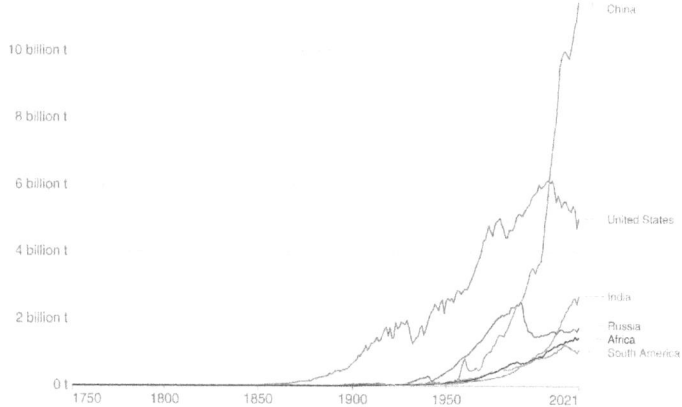

The US has generated more than its fair share of carbon emissions over the years (from Our World in Data)

This, in turn, leads to some actual science: how does the greenhouse effect work? And these guys don't even know what a greenhouse is! What are the sources of greenhouse gases? Are some power sources better than others, in terms of carbon emitted? How much power does the U.S. get from each kind of power source? Some of these are appropriate for a pie chart—and they love pie charts, you can do so much visually in them. The data contain some eye-openers for them, and for me, as well. Here's one: look at a graph of carbon emissions by country over time. Right now, China leads in emissions, by a lot. But until 2005, it was us. Indeed, the

Tim Coonan

good ole USA has produced the most CO2 emissions over time, almost 400 billion tons since 1750. That's more than twice as much as China has produced. You know what, kids? This problem is ours to solve, isn't it? We made the mess; now we need to clean it up. The US has one of the highest per capita rates of carbon emission as well. Makes one think twice about driving a car; our lifestyle may be dooming low-lying island nations, for one. Not quite fair, is it?

The other thing I desperately want them to know is that climate change is not an abstract, some-day-in-the-future, maybe-it-will-affect-your-grandkids thing. It's happening right now. In fact, it's already happened. I assign them each a natural or cultural resource to research, something of value that has already been affected by climate change, and I ask them to do something they do so well, something that, in fact, may be their favorite type of assignment. I ask them to make a slide presentation. They are so good at this! The array of visual images, with special effects transitions! Veiled references to fifth grader inside jokes and popular culture! And I find that when kids teach each other, retention and interest and engagement are the highest. And the kids absolutely nail it. Ayden M. shows us predicted map images of Venice being all but swallowed by the sea; Gabrielle does the same for Florida, which may lose as much as two-thirds of its land to the rising ocean. Ezekiel shows us the current and predicted decline of polar ice. The effect of polar ice decline on polar bears, indeed, the predicted extinction of same, a now-familiar poster child for climate change, is presented by Ayden F., who has a place in his heart for animals (and is still obsessed with dinosaurs—he knows so much more about them than I do!). Andrew details the loss of ice sheets and glaciers, Gabe the projected decline of the Amazon rain forest, and Noah tells of the widespread effect on agriculture. And Kate—bless her heart! Kate, whose mom would die later that year of an aneurism, whose dad would end up taking the kids back to Ireland, Kate researches the effect of climate change on, as she calls it, forest fires. So we all learn, together, what Kate has researched: that rising temperatures have given us longer fire seasons, drier fuels, hotter, bigger and more frequent fires. And those fires have spewed even more carbon into the air. Nicely done, Kate.

We spend some time—but just some—on the climate models and whether we will meet our climate goals: enough decrease in carbon emissions to keep global temperature rise at 1.5° or 2.0°C. It's easy enough for adults to get lost in the modeling and the projections, the minutiae and political exigencies of meeting climate goals, let alone kids. And climate justice? That's another thing altogether. At some point, it will be important for them to think about the irony of asking developing nations to slow their development, to do it nicely, so we can continue to live our developed world lives. Someday they also need to know that communities of color will fare—are faring—worse than white communities here in the U.S., where a history of redlining, of racist development policies, has relegated those communities to areas with higher heat loads and less respite from the effects of climate change.

But all that can wait. Because what I also want them to know is that there is some hope out there. And not just the hope that THEY themselves represent. They will be the generation that implements our solutions to the climate change puzzle, and lives to see whether we did okay. It's been inspiring to see the commitment the younger generation has to fighting climate change and insuring climate justice. Look at what Greta Thumberg, in her short life, has accomplished. And I personally need look no farther than my own daughter. Carrie, who has marched for climate change, has committed herself to a vegan diet, for environmental reasons, and it's hard to argue with that: cattle are a huge source of methane, one of the worst of the greenhouse gases, and ranching practices are detrimental to public lands and to carbon balance. I, myself, don't really know what to do with tofu (it's a medi-

Tim Coonan

um in which I don't work), but Carrie sure does. I'll let her fix tofu for me any time.

Tofu aside, here's some hope that the fifth graders can understand. See, they already know about the relative contribution of different power sources to U.S. energy, because I asked them to graph it: natural gas about 30%, the same for petroleum; about 10% each from coal, renewables, and nuclear. All right, guys, find a country that's doing this right, with more renewables and less from coal and natural gas. And they do! They are adept at quick research on the Internet (Sources, guys! Document your sources!). There are countries out there who are absolutely doing this right. Thanks to Noah, we find out that Scotland gets almost all of its energy from renewables, and 70% of that from offshore wind. Iceland, too, does quite well: Ayden M. reports that Iceland also gets most its energy from renewables: about a quarter from geothermal, and the remainder from hydro, taking advantage of its natural resources. The kids see that we are not at all locked into reliance on fossil fuels, that we can—indeed, must—tweak our sources, break the iron grip that fossil fuel development has long held on us, a grip that has heretofore doomed us to a planet more hostile, more inimical to life. A Blue Planet turning red, as it were (and that analogy is just heat-based, and not political; but if the shoe fits...).

It also occurs to me that these kids, whose professional careers lie a decade or so in the future, may work on these climate problems themselves, in ways that I, a conservation professional of 30+ years, never could have imagined. When I entered the workforce, the opportunities to make a difference in conservation were limited to being a conservation scientist (working on the natural resources under threat by human actions), working for an advocacy group, such as the Sierra Club, working for a land management agency, such as NPS or USFS, or being an environmental lawyer. I thought the latter had enormous potential for bringing about environmental change. So much so that in my last year of grad

Lake Effect

school I took a step down that road: I took the LSAT—even got a free pencil out of it, a green one that said "LSAT" on it. But then I realized that I'd mainly be working indoors. That was a dealbreaker for me; I just wanted to be OUT THERE. Wildlife biologist it was.

But the opportunities to work on environmental issues, including combatting climate change, have just exploded in recent years. Solar power, wind power...those alternative energy sources are more and more mainstream these days, are an accepted part of our power grid, as fossil fuels are put on a back burner (maybe on an energy-efficient induction stovetop). I used to see solar power and wind power arrays as a blight on the desert landscape; now I love seeing them. The electrification of our vehicle fleet—what an enormous task! And one rife with possibilities. My partner Nell works for an electric vehicle charging company ("That's righteous work," my friend Diane said when introduced to her). Her company has expanded tremendously as they, and other charging companies, attempt to expand the nation's network of vehicle charging stations, funded and supported by governments, power companies and car manufacturers. Everyone knows which way the wind is blowing on this one.

You can take this a step further, and this seems almost well, sly to me. Nell's daughter Katie is a college senior, majoring in environmental studies, but with a business minor. Investment banking. She spent her summer before senior year in New York City, as an investment banking intern for a large U.S. bank. Her work involved convincing companies to invest a greater share of resources in renewable energy. She plans to do this, and change the world for the better through investment banking. Katie, do they KNOW this is what you want to do? And they're okay with it? You're not pulling the wool over anyone's eyes, are you? It's like having a spy on the inside, across enemy lines. Change the world through investment banking. I'm all for it! There's a lot of righteous work out there. Just don't forget to go hiking on

Tim Coonan

the weekends.

Chapter 21

The Otter, the Nutcracker and the Wolf

On our way to Almanor in 2022, Nell and I stopped to buy sheets and towels at a Target in Chico. For as many years as I'd been doing this, that year I had forgotten to bring linens, and thank goodness Targets are ubiquitous. It was 106° in Chico. There was no shade in the Target parking lot; I hoped my paddleboard, lashed to the top of the car, wouldn't delaminate, just melt all over the car, like buttercream frosting in the sun, while we were in the store. Later that summer a record-breaking heat wave would engulf the entire U.S.—I had never seen the weather map so red—and if you ever doubted that climate change was real, all you had to do was step outside. A European heat wave drew down that continent's rivers to the point where Rhine and Danube river cruises were shuttled onto buses. On this continent, reasonable scientists, usually not prone to hyperbole or alarmism, concluded that Lake Mead and Lake Powell would never be full again, and would likely soon approach "deadpool" status —levels so low that power could no longer be generated from their massive dams. Climate change was staring us right in the face.

I was anxious, apprehensive, as we wound our way up out of Chico toward the lake, for several reasons. First, I really didn't know what I'd find, after the Dixie Fire's devastating run the summer before. A burnt umber moonscape? Mordor? Hot ashes for trees? Hot air for a cool breeze? That was a million-acre fire, for chrissakes, and it had charged up both sides of the lake. How much would our Almanor experience be affected by the fire which had chased us out the year before?

And then there was Nell. I really, REALLY wanted her to like the lake, which was such a significant place for me, and my family, despite being, in reality, just a smallish, manmade reservoir among the pines. And Nell was a fresh set of eyes, an objective, outside perspective on what may just have been a rose-colored and nostalgic experience for me.

We climbed up out of Chico, and the housing developments on the outskirts of town gave way to dry grasslands, fields bordered by dry volcanic rock walls, oddly reminiscent of the rock walls demarcating fields in Ireland and Scotland. The same volcanic rocks studded the fields. Even down here, the volcanic history of the area was apparent, was on display. We drove through dry, hot, scrubby pines, and finally the cool pine forest, and then the road plunged down Deer Creek. It was the same: a gurgling stream overhung by sycamores, alders, willows, grading into the pine forest. Forest Service campgrounds. Fishing spots, swimming holes. And at the bottom, wet meadows lined by conifers and split-rail fences. The fire had not penetrated this lush corridor, this Rivendell, protected, if not by elven magic, then by its wet nature, or perhaps merely by happenstance, the vagaries of wind and topography which govern fire spread. Whatever. I'll take it. The grand entrance to Lake Almanor was intact.

As we turned east on Highway 36 toward Chester and the lake, last year's fire began to make its presence known. We saw logging trucks coming the other way, their loads blackened and charred. On the outskirts of Chester, the fire, on its run up above the western shore, had blown through a swath of conifer, miles wide. The trees were still standing; they too were blackened and charred, their needles brown. Not a green tree in sight. These were dead, and not wholly consumed by the fire. They were a hundred or more feet high, and I could only imagine the flame lengths as the Dixie express had roared through on its way to Lassen. The Forest

Service—I assume—had cleared a swath of these burned trees back from the road itself; this was the source of the charred logs on passing trucks. It looked devastated, and yes, a bit like Mordor. We drove into Chester, and it was how I remembered it, saved by the lake on one side and what I thought had been aggressive actions such as back-burning on the other. Our friend Joe later told us only a shift in the wind had saved Chester.

Cattle graze in front of charred "matchsticks", outside of Chester.

We pulled out of Chester, over the causeway that bridges the marshy area, dotted now with Canada geese and great egrets, and even white pelicans, those inland cousins of my southern California-based brown pelicans. The white pelicans are denizens of Western lakes, well-known for breeding on Yellowstone Lake, for example, and it was great to see them at Lake Almanor. The pine forest was thick and intact as we headed uphill toward the head of the peninsula. It reminded me of something I had forgotten—that wildland fires are patchy, flighty things, almost randomly leaving some areas whole and untouched, while just devastating others, not really by whim but by combinations of wind, weather and topography which we may understand a bit (fire science being a thing) but over which we have no control at all. So the view east from the peninsula was deceptive. There was a burned swath visible toward the top of Dyer Mountain (the fire had passed on its eastern slopes), but otherwise unsullied conifer forest stretched as far as you could see. To the southwest and west it was a different story. The hills beyond the lake shore itself were, as Joe said, matchsticks. And in the view to the northwest, in front of Lassen's bare talus slope, with its receding patch of glacier, you could see the swath of "matchsticks" left behind.

We had a great week at the lake. It was a smaller crew—just Dan, Nora and Dennis' families, half of the previous year's crowd. Nell and I met them at the concert at the park that night, and there was even an Eagles tribute band, which boded well for the week. Dan had rented us a house on the eastern shore of the peninsula (the previous year we had been on the western side of the peninsula, with its view of Lassen) so we looked out on Dyer Mountain, and sunrise. Nell and I were up before dawn the next day. In fact, every day we got up before dawn—we don't want to miss a sunrise, and we like to be doing something in that first light. We took our coffee down to the dock. Grebes were out on the lake, intermittently calling to each other and diving for fish. A family of mergansers effortlessly floated by, close to shore, and I'm sure their webbed feet were paddling furiously underwater. I saw a strange, humped back break the surface of the lake. My southern California eyes briefly thought it was a sea lion; it was definitely a mammal. A river otter! It surfaced again and headed in toward the docks to hunt for fish. The enmity the local fishermen have for the otters makes me laugh. I don't imagine they take THAT many fish, although they apparently do a number on the grebes during breeding season. Hey, more power to 'em. Just taking advantage of this relatively new lake, like the rest of us.

We did other lake stuff. Went down to the tip of the

peninsula, with its 270° view of the lake, including Lassen. My dad loved going down there, after Mom passed. It gave him—as it gives me—a sense of perspective on this place, an opportunity to remember ALL of it: horseback riding on the west shore, patio boat trips to Coonan Cover, milkshakes and Mass in Chester. Nell and I passed the Hillyer place, still apparently owned by the Hillyers, their carved sign out near the road; she had heard the stories. "Is this the place with the firepole?" she asked. Yup. We saw does and tiny, speckled deer fawns on the lawns, and they delighted Nell as they had delighted us citified kids years ago (and still do).

We ate pizza and drank Moscow mules at Tantardino's, which in another sign of change I find hard to accept, is not Tantardino's anymore. The Tantardino sisters had sold the restaurant. It is now called Il Lago (Italian for "The Lake"), and luckily the menu, and apparently much of the kitchen staff, is pretty much exactly the same. We ate there three times during the week. Across the street from the restaurant sits the local brewery, with a wonderful Maidu name, Waganupa, which no doubt means "hazy IPA" or "pilsner" or something similar in that ancient language. The brewers there have taken advantage of the Dixie Fire, made lemonade when handed lemons, so to speak. The Dixie Fire put fine particulates in the air, as fires tend to do, and that affected the yeast fermentation process: the resulting beer is called "Dixie Dust." Maybe it has a hint of pine tree in it.

Dock Talk, of course. Nell was initiated into the music selection (you only get ONE song at a time). A couple nights we stayed pretty late on the dock, and dinner was indeed delayed, until about 10:30 or so. We celebrated Nora and Neal's 30th wedding anniversary on the dock, adding that to the several graduations that have been dock-celebrated (Dan holds high hopes for a family wedding on the dock, or at least a marriage proposal). One night, while we were on the dock, a striped skunk ambled by on shore. Actually, it was moving pretty fast, trotting along, looking for god knows what, whatever skunks like to eat (and they are little carnivores). For a brief moment we thought it would make a turn and trot out the gangplank to the dock, where we were. Joe had a plan: leave the phones on the dock and jump in the water. Seemed reasonable to me. I like skunks, because for one, they are black and white, and I find that attractive and unique in an animal. Zebras. Orcas. Pandas. The other thing I like about skunks is that that they, like jays, really don't give a fuck. They just do what they want to do, and they have the scent glands to back that up. The other species that absolutely does not give a fuck is the bear. They do EXACTLY what they want to. I love that. Joe and Michelle told us more bears had been seen that summer on the peninsula, going through trash cans at night; I imagine that changes wrought by the Dixie Fire had something to do with that.

That night the bats appeared, like clockwork, right at their magic hour, when dusk has fallen, but before darkness quenched everything but the stars and the crescent moon. Nell, again, was enchanted. The bats swooped over the water around the dock and out to the moorings, dozens and dozens of them, rollercoasting up and down mere inches above the water. Hitting on the bugs that were also out at that hour; I imagine the fish were hitting them, too. The window of opportunity is relatively short, the bats feeding, at least in that area, for less than half an hour. Carrie wasn't there to cruise slowly among them, on a paddleboard, but I marveled at the spectacle, all the same. It seems to me that natural selection has produced compromises, in many cases. Here, the insects are out at dusk, because they would be easy prey for fish in the full light of day. So the bats hit them up at dusk, using their excellent echolocation, rather than their somewhat marginal sight, to find them. But the insects also are out at dawn, the other crepuscular period, as any angler knows, and that's when the bats' daytime counterparts, the swallows, hit on them. And the grebes and the ospreys feed on the fish at these hours, too. I observed all this Mutual of

Lake Effect

Omaha Wild Kingdom stuff the next morning, on our dawn paddle (well, I was paddling and Nell, the triathlete, was getting her long swims in). I slowly paddled up the peninsula and Nell followed me. The pace allowed me to observe everything in the quiet morning scene. Swallows were swooping up and down on the calm waters near the Rec 1 cove. Just past that cove, an osprey sat in a nest at the top of snag; its mate swooped over us, looking for fish in the calm morning waters, no doubt, but I also sensed it was telling me to move along. These were its waters, and its territory. No problem, man, not here for your fish, or for your babies.

The burned trees were a trail of breadcrumbs to Lassen Peak, as if that had been the fire's military objective all along.

After the paddle/swim Nell and I brought coffee out to the deck, and Dan joined us in this favorite of my Almanor morning rituals. Few boats disturbed the calm of the lake, just a couple of slow-moving anglers trolling the shallows, and truth be told, relatively few boats would be seen on the lake throughout the day; it is just not that busy, compared to Tahoe or Shasta. The lake and Dyer Mountain were framed by the tall conifers, and right near the deck were mature specimens of the Big Three: ponderosa pine, white fir, and incense cedar. I could lean my head back and look at the

Tim Coonan

foliage, the three needle types distinct, disappearing into the blue sky. The forest inhabitants showed up on their morning rounds. Raucous Steller's jays (not giving a fuck), and the gleaners, the mixed foraging flock of small birds probing the boles and branches for insects. Mountain chickadees. Black-headed juncos. Nuthatches, their presence announced, subtly, by their soft honks. One landed on the trunk of the pine and made its way down, probing away. "What's that bird?" asked Dan. "It's a red-breasted nuthatch," I replied. "You're just making that up," he joked. The sun rose a little higher over Dyer, and the day on the lake was pure potential.

Midweek Nell and I went to Lassen; I had to climb that peak again, and I had to see what the fire had done to that park. I had also recently viewed the "historic" footage of my dad's from our 1971 trip to the park, and wanted to check those areas out: Sulphur Works, Bumpass Hell, Ridge Lakes. The road into Lassen cuts up from Highway 36, which in that area was fairly untouched by the fire, but the park road itself ran past miles of burn; the park had been closed for quite some time after Dixie came and went. Trees lay dead or still stood, and Nell asked if the National Park Service would remove them and replant. That's not really how the NPS works, I told her. Trees would eventually sprout in the ash-rich soil, and a new generation of tree saplings would follow the forbs and shrubs, taking advantage of the rich soil and open canopy. The particular tree species would depend on the elevation of each site, and the mix might be slightly different from what was there before, given rising temperatures and other changes, such as less snow and soil moisture, brought about by climate change. But I was pretty sure there'd be no salvage logging of the burned-over sites. It's just not what the NPS does.

The road climbed in elevation, leaving the burned areas, and we passed Emerald Lake and Lake Helen, beautiful glacier-carved lakes in the Lassen high country. The lakes had the glacial coloration, especially when seen from above: deep

Lake Effect

blue interiors and green edges. The trail to Lassen Peak starts just beyond Lake Helen, and as I observed, you get a lot for your money with this hike. It's only five miles round trip, but you get to bag a peak which is above 10,000 feet, with only 2,000 feet of elevation gain. There were relatively few people on the trail, quite a contrast to our ascent of southern California's Mt. Baldy the week before (which was twice the distance, twice the elevation gain, and had five times as many hikers). Yes, Lassen is a hidden gem of the National Park system. Because the trail starts at above 8,000 feet, you're immediately in the whitebark pine forest, which, as it approaches timberline, grows sparser, and the trees become crumpled and wind-sculpted. We saw Clark's nutcrackers in the pines; they have a great relationship with the whitebarks. The jays probe their cones for seeds, which they cache, burying them for future consumption. Luckily for the pines, the birds can't remember where they cache ALL the seeds (and each nutcracker might bury tens of thousands of seeds each year), and so some of those seeds sprout, jump-starting the next generation of whitebark pines.

The view from Lassen's peak is still deceptively benign, belying the trouble below.

Tim Coonan

The trail above timberline switchbacks up to the side of bare talus slopes, and the view gets better and better, the wind threatening to unhat you at every western mountain face. You reach the ridge top and Mount Shasta looms on the horizon, as it should—its peak is above 13,000 feet—another reminder of the volcanic origins of everything around you. As the Park Service says, every rock you see at Lassen is volcanic in origin. Up here you also get a good feel for the massive volcano, Mount Tehama, that existed here 500,000 years ago; its crater was eleven miles across, but all that's left is Brokeoff Mountain to the southwest, now just shards of its original cone. Mount Tehama certainly dwarfs the much more recent (27,000 years old) and smaller Lassen Peak.

We crossed a snowfield (Nell stopped to throw a snowball at me) to get to the actual summit, a jagged jumble of rocks populated by fellow hikers quietly eating their lunch and trading off taking each other's' pictures. The land dropped precipitously in every direction, and Lake Almanor looked pretty small in the distance, but its heart shape was instantly recognizable. Funny how everything looks pretty good from up high. Remarkably benign, like those shots of Earth, our Blue Planet, from the Moon. You could still see some matchsticks, of course, and there was less snow than when Bridget and I hiked up here in 2016, and there was FAR less snow on Lassen than in the footage my dad shot in 1971. But it's relatively easy to forget all that when you're up that high. I wondered if, in future visits, the summer snow would disappear altogether from Lassen's slopes and top, as rising temperatures and less snow reduce summer snowpack. The rising temperatures will force some species up in elevation, but there's a limit to that, since there's considerably less area at higher elevations than at lower ones; mountains slim down as they go up. Species like the pika, the so-called farmer of the tundra, may be squeezed to extinction, might just run out of suitable habitat. All this takes some effort to think

Lake Effect

about, to confront and accept, when you're under the physical and emotional spell of a high-elevation, peak-bagging high.

It was a wholly pleasant hike back down, past Lassen's steep talus slopes, scalloped and sculpted by the last of the Pleistocene glaciers. We saw no pikas, or marmots, but I knew they were there, still, not yet disappeared by climate change. Perhaps in the future, but not yet. We stopped at Bumpass Hell (I still love that name) and Sulphur Works on the way back. I thought of images from my dad's home movies: my mom pushing Dennis in a stroller along these paths in 1971, four-year old Nora bouncing up the path, blonde pigtails and all, to meet them. We boys hiking down from Ridge Lakes; I'm pretty sure the trail sign there now, at the beginning of the trail, is the same routed wood sign from 50+ years ago in my dad's movies. A Park Service standard, those routed wood signs are, as recognizable and emblematic as the NPS arrowhead and the wonderful lettering, that old-timey Art Deco-ish font, on the vintage National Park posters. These volcanic features, Bumpass Hell and Sulphur Works, are much the same as they were fifty years ago. Here the magma, the molten rock, occurs so close to the surface that groundwater bubbles up into mud pots and steam vents, the magma being a seemingly endless source of energy that has not abated in the interim, in those fifty years, during which so much has changed in my life and in the country, the world at large. Lassen is still a wonderful song of ice and fire, where the forces of volcanism and glaciation are on full display, the former still active, the specter of future eruptions unpredictable but assured; the latter force now past, evident in the glacial lakes and carved topography.

We ended our visit at, appropriately enough, the park's visitor center, where I learned the Dixie Fire had burned through 70% of the park, and park staff had collated comments from park visitors on what they saw, and what it meant to them. All the visitors were saddened by what they saw, and why wouldn't they be? The national parks we visit are like old friends; we don't want them to change; and while I fully accept the dynamic nature of, well, nature, I don't want these old friends to be, as the NPS says, impaired. And these climate change-induced changes are indeed impairments; the catastrophic fires imperiling sequoias and the very character of our Western forests are so far beyond the pale, so far outside even the observed wide variation in natural conditions, that parks may change forever. Irreversible changes, due to human actions. The most we can hope to do is limit the damage, slow down the speed so the accident isn't fatal. The comments I saw posted, some from kids, were a lament, an elegy for the West that we have known, have grown up with, are growing up with.

The rest of the week? You could say it was a quiet week at Lake Woebegone - er, Lake Almanor. Especially the last night; everyone else left a day early, and Nell and I had the place to ourselves, which was just unprecedented. We nonetheless did everything a Lake Almanor vacation entails. Swam and paddled in the early morning, had coffee on the deck, lay in the sun on the dock. Went to the brewery, ate at Tantardino's (I mean, at Il Lago). Even went back and had our own Dock Talk, with the music selection alternating between the two of us. The bats appeared at their appointed hour, and the otter showed up as well, maybe to say goodbye.

I was dying to know what Nell thought of the week, of the lake, but I didn't press her. Didn't want to jinx it. But that last night on the dock, Nell turned to me and said, "You know, I had low expectations about Lake Almanor. But I loved it." She loved all of it! She had grown up in upstate New York, on one of the Finger Lakes; had summered in Maine, and had suitably high lake vacation standards because of it. Heretofore, her California lake experiences had been underwhelming, at best, limited to the crowded shorelines and drunken boater traffic at southern California's Big Bear

Lake Effect

Lake. She expected nothing more from Almanor. But the quiet, the relative lack of boaters, the sense of space, the mergansers, geese, grebes, ospreys, otter and bats won her over. She loved swimming in the lake, and even appeared to enjoy our own brand of dock-centered lake vacation.

Dock Talk for two (unless you count the bats).

It's hard to say what the future holds for Lake Almanor. Certainly wildfires will encroach on the lake, and our future vacations there; it was just a matter of luck that they didn't in 2022. The other effects of climate change will be more insidious. Slow and inexorable. Will climate change bring Almanor's water level down, the way drought has dropped other reservoirs in the West? Lake Mead and Lake Powell, are only half full at this point, and will likely never be full again. Almanor? It is just a reservoir, after all. Lake Oroville, just downstream from Almanor and the largest reservoir in California, was only half-full in 2022, as well. How long

Tim Coonan

before those effects move upstream, up the North Fork of the Feather River, and force those Almanor docks far away from shore? Hard to imagine that the lake of my childhood and adult vacations would be drawn down to the point of being unrecognizable. That temperatures would rise to the point where there is no summer snow on Lassen Peak at all, and pikas no longer live on its rocky slopes. And this silly vision I have of Almanor, of reservoir as respite, turns out to be just pure wishful thinking on my part, fantasy, nothing more than nostalgia, a hearkening back to the happy days of my youth. Perhaps, in the end, it is just true what Virgil said, quoted by Willa Cather: *Optima dies...prima fugit.* The best days are the first to flee.

But life is tenacious, isn't it? Resilient. It not only finds a way, it desperately claws its way out of a hole and then thrives. Survival of the fittest, indeed. In August 2021, as the Dixie Fire had reached the 850,000-acre mark and was not yet done, California's wolf biologist went looking for the Lassen wolf pack, at their last known radiolocation. The area was just matchsticks. Dead cattle, cows and calves. The biologist expected dead wolves, as well. But in a lush clearing, spared by the fire, he heard a low growl. The pack's alpha male and female, as well as four pups, had survived, had weathered the blaze, God knows how. The biologist called it total luck, and that certainly may have had something to do with it. But perhaps it was more than that. Perhaps the adults had been smart enough, savvy enough, experienced enough, to lead their pups to safety. Certainly a sign of hope, isn't it? Maybe, just maybe, we, as a species, can do the same, be smart enough to see our way through this. I certainly hope so. I want to see that otter every time I go to Almanor, and nutcrackers on the slopes of Lassen. I want those wolves to be out there still, somewhere beyond Dyer Mountain, doing wolf things.

About the Author

Tim Coonan is a wildlife biologist who spent 30 years with the National Park Service, most notably at Channel Islands National Park, where he led a successful recovery program for an endangered species, the island fox. Tim then taught science to impressionable middle-schoolers, which in some ways were more difficult to work with than island foxes. Tim grew up in suburban Los Angeles, graduated from Notre Dame, and earned a masters at Northern Arizona University. He co-authored the definitive (granted, the only) book on the island fox, and published a memoir about family station wagon road trips (Our Lady, Queen of the Highways) in 2022. He has two grown daughters and lives in Ventura, where he enjoys living a mostly-outdoor life (you know, hiking and all that) with his partner, Nell, who also runs triathlons. Tim watches Nell run triathlons.